THINKING STRATEGICALLY

THE APPROPRIATE USE OF METRICS FOR THE CLIMATE CHANGE SCIENCE PROGRAM

Committee on Metrics for Global Change Research
Climate Research Committee
Board on Atmospheric Sciences and Climate
Division on Earth and Life Studies

NATIONAL RESEARCH COUNCIL
OF THE NATIONAL ACADEMIES

THE NATIONAL ACADEMIES PRESS
Washington, D.C.
www.nap.edu

THE NATIONAL ACADEMIES PRESS • 500 Fifth Street, N.W. • Washington, DC 20001

NOTICE: The project that is the subject of this report was approved by the Governing Board of the National Research Council, whose members are drawn from the councils of the National Academy of Sciences, the National Academy of Engineering, and the Institute of Medicine. The members of the committee responsible for the report were chosen for their special competences and with regard for appropriate balance.

This study was supported by the federal agencies of the U.S. Climate Change Science Program through the National Aeronautics and Space Administration under Contract No. NASW-01008. Any opinions, findings, conclusions, or recommendations expressed in this publication are those of the author(s) and do not necessarily reflect the views of the organizations or agencies that provided support for the project.

International Standard Book Number (ISBN) 0-309-09659-6 (Book)
Library of Congress Control Number 2005929740

Additional copies of this report are available from the National Academies Press, 500 Fifth Street, N.W., Lockbox 285, Washington, DC 20055; (800) 624-6242 or (202) 334-3313; Internet http://www.nap.edu

Cover design by Van Nguyen, the National Academies Press.

Printed in the United States of America.

THE NATIONAL ACADEMIES
Advisers to the Nation on Science, Engineering, and Medicine

The **National Academy of Sciences** is a private, nonprofit, self-perpetuating society of distinguished scholars engaged in scientific and engineering research, dedicated to the furtherance of science and technology and to their use for the general welfare. Upon the authority of the charter granted to it by the Congress in 1863, the Academy has a mandate that requires it to advise the federal government on scientific and technical matters. Dr. Ralph J. Cicerone is president of the National Academy of Sciences.

The **National Academy of Engineering** was established in 1964, under the charter of the National Academy of Sciences, as a parallel organization of outstanding engineers. It is autonomous in its administration and in the selection of its members, sharing with the National Academy of Sciences the responsibility for advising the federal government. The National Academy of Engineering also sponsors engineering programs aimed at meeting national needs, encourages education and research, and recognizes the superior achievements of engineers. Dr. Wm. A. Wulf is president of the National Academy of Engineering.

The **Institute of Medicine** was established in 1970 by the National Academy of Sciences to secure the services of eminent members of appropriate professions in the examination of policy matters pertaining to the health of the public. The Institute acts under the responsibility given to the National Academy of Sciences by its congressional charter to be an adviser to the federal government and, upon its own initiative, to identify issues of medical care, research, and education. Dr. Harvey V. Fineberg is president of the Institute of Medicine.

The **National Research Council** was organized by the National Academy of Sciences in 1916 to associate the broad community of science and technology with the Academy's purposes of furthering knowledge and advising the federal government. Functioning in accordance with general policies determined by the Academy, the Council has become the principal operating agency of both the National Academy of Sciences and the National Academy of Engineering in providing services to the government, the public, and the scientific and engineering communities. The Council is administered jointly by both Academies and the Institute of Medicine. Dr. Ralph J. Cicerone and Dr. Wm. A. Wulf are chair and vice chair, respectively, of the National Research Council.

www.national-academies.org

COMMITTEE ON METRICS FOR GLOBAL CHANGE RESEARCH

CLIMATE RESEARCH COMMITTEE

Acknowledgments

This report has been reviewed by individuals chosen for their diverse perspectives and technical expertise, in accordance with procedures approved by the National Research Council's Report Review Committee. The purpose of this independent review is to provide candid and critical comments that will assist the institution in making its published report as sound as possible and to ensure that the report meets institutional standards for objectivity, evidence, and responsiveness to the study charge. The review comments and draft manuscript remain confidential to protect the integrity of the deliberative process. We wish to thank the following individuals for their review of this report:

Jack Azar, XEROX Corporation, Webster, New York
Rosina Bierbaum, University of Michigan, Ann Arbor
Susan Cozzens, Georgia Institute of Technology, Atlanta
Jack Fellows, University Corporation for Atmospheric Research, Boulder, Colorado
Debra Knopman, RAND Corporation, Arlington, Virginia
Roger Lukas, University of Hawaii, Honolulu
Michael Mann, University of Virginia, Charlottsville
Philip Marcus, University of California, Berkeley
Linda Mearns, National Center for Atmospheric Research, Boulder, Colorado
Elinor Ostrom, Indiana University, Bloomington
David Skole, Michigan State University, East Lansing

Although the individuals listed above have provided many constructive comments and suggestions, they were not asked to endorse the conclusions and recommendations nor did they see the final draft of the report before its release. The review of this report was overseen by Robert Frosch, Harvard University, and Thomas Graedel, Yale University. Appointed by the National Research Council, they were responsible for making certain that an independent examination of this report was carried out in accordance with institutional procedures and that all review comments were carefully considered. Responsibility for the final content of this report rests entirely with the authoring committee and the institution.

Preface

Federal agencies are increasingly being asked to document progress and measure performance to improve their accountability to Congress and the public and to provide information useful for making budget decisions. This task can be difficult to accomplish, especially in a program as complex as the Climate Change Science Program (CCSP), which spans all of the environmental and related social science disciplines and includes activities ranging from basic research to decision making in 13 federal agencies. Current approaches to evaluate progress (e.g., peer review of basic research, reduction of uncertainty) focus on particular aspects of the CCSP and/or have other limitations. For example, gaining improved understanding of the climate system can lead to increased uncertainties about some aspects of the system, yet progress has clearly been made. So, the question remains: How can progress in climate science be demonstrated after nearly 15 years of sponsored research and observations?

At the request of Dr. James Mahoney, director of the U.S. Climate Change Science Program and chair of the Subcommittee on Global Change Research, the National Research Council established a committee to develop quantitative metrics and performance measures for documenting progress and evaluating future performance for selected areas of global change and climate change research. Committee membership included researchers drawn from a wide range of global change disciplines and experts from industry and government with practical experience in developing and using metrics.

The Committee on Metrics for Global Change Research held three meetings from December 2003 to June 2004 to discuss the issues and to gather input in three major areas:

1. the different types of metrics (e.g., input, outcome) and the different scales of programs that can be evaluated usefully by such measures;
2. the experience of industry, academia, and federal government agencies in measuring performance; and
3. lessons learned from retrospective analysis of science programs.

A fourth meeting (September-October 2004) was devoted to writing this report. In preparing its report the committee strove to provide practical advice on the applicability of performance measures across the full range of CCSP goals and approaches—from discovery science, to modeling and assessment, to communicating results and managing risk.

The committee thanks the following individuals for making presentations or providing other input: David Bader, Susan Cozzens, James Hack, Richard Hallgren, Jack Kaye, Charles Kennel, Mike MacCracken, James Mahoney, Richard Moss, Franklin Nutter, Cheryl Oros, John Parascandola, Craig Robinson, Jason Rothenberg, Sherwood Rowland, and Spencer Weart. Thanks also go to members of the Climate Research Committee and Board on Atmospheric Sciences and Climate—particularly Anthony Busalacchi, James Coakley, David Karoly, Robert Serafin, and Lynne Talley—for their input and guidance throughout the study. Finally, the committee extends its appreciation to the NRC staff, particularly study director Anne Linn, for their highly professional contributions to this report.

Eric Barron
Chair

Contents

Executive Summary

*T*he Climate Change Science Program (CCSP) and its predecessor U.S. Global Change Research Program (USGCRP) have sponsored climate research and observations for nearly 15 years. Although significant scientific discoveries and societally beneficial applications have resulted from these programs, the overall progress of the program has not been measured systematically. Metrics—a system of measurement that includes the item being measured, the unit of measurement, and the value of the unit—offer a tool for measuring such progress, improving program performance, and demonstrating program successes to Congress, the Office of Management and Budget, and the public.

Metrics have been applied successfully to research programs in industry, academia, and the government. The challenge is applying them to a complex program such as the CCSP, which involves 13 federal agencies and sponsors a wide range of activities—from basic research on the earth-ocean-atmosphere-human system, to assessment and risk analysis, to decision making. At the request of the James Mahoney, director of the Climate Change Science Program and chair of the Subcommittee on Global Change Research, the National Research Council's Committee on Metrics for Global Change Research was convened to

1. provide a general assessment of how well CCSP objectives lend themselves to quantitative metrics;

2. identify three to five areas of climate change and global change research that can and should be evaluated through quantitative performance measures;

3. for these areas, recommend specific metrics for documenting progress, measuring future performance (such as skill scores, correspondence across models, correspondence with observations), and communicating levels of performance; and

4. discuss possible limitations of quantitative performance measures for other areas of climate change and global change research.

The committee approached its task first by examining the experience of industry, federal agencies, and academia with implementing metrics, and then by formulating possible metrics for a wide range of CCSP objectives. It began its deliberations with some skepticism as to whether metrics would apply to many of the elements of the program. However, analysis showed that it is possible to develop meaningful and useful measures for all parts of the CCSP. The difficulty arises in selecting a few areas of global change and climate change for which metrics should be developed (charge 2). The committee found that it was not possible to make this selection without a clearer sense of program priorities. The CCSP strategic plan does not contain measures of success, and program objectives are written too broadly for them to be inferred. However, even if such guidance were available, the committee found that a broader range of quantitative and qualitative metrics would be a more valuable tool for managing the program. The key to promoting successful outcomes is to consider the program from end to end, starting with program processes (e.g., planning and peer review) and inputs (e.g., resources) and extending to outputs (e.g., assessments, forecasts), outcomes (e.g., results for science and society), and long-term impacts. Principles and a framework for creating and implementing metrics for the entire CCSP are described below.

PRINCIPLES FOR DEVELOPING METRICS

Industry, federal agencies, and academia have different objectives in developing metrics. Industry has long used metrics to gauge progress in meeting business objectives and to identify where adjustments should be made to optimize performance and increase profits. Federal agencies are increasingly relying on metrics, either to manage programs or to increase their accountability to Congress and the public. The latter motivation was strengthened by the Government Performance and Results Act of 1993, which required federal agencies to set strategic goals and to measure program performance against those goals. Finally, academia uses metrics to supplement peer evaluation in decisions to hire or promote faculty members, allocate resources among departments, or compare the performance of departments at different universities.

Based on the collective experience of these three sectors, the committee offers the following principles for developing useful metrics and avoiding unintended consequences:

1. Good leadership is required if programs are to evolve toward successful outcomes. The overall program will suffer if no one has the authority to direct resources and/or research effort and to develop and apply metrics. The leadership of a few individuals in supporting research and/or publicizing the implications of research results, for example, helped speed understanding of the causes of Antarctic ozone loss. These actions ultimately led to regulations on the reduction of chlorofluorocarbon emissions, which are expected to return effective chlorine amounts in the stratosphere to pre-ozone-hole conditions by mid century.

2. A good strategic plan must precede the development of metrics. Such a plan includes well-articulated goals against which to measure progress and a sense of priorities. Absent this context, it is difficult to select the most important measures for guiding the program.

3. Good metrics should promote strategic analysis. Demands for higher levels of accuracy and specificity, more frequent reporting, and larger numbers of measures than are needed to improve performance can result in diminishing returns and escalating costs. The nearly continuous assessments of the Intergovernmental Panel on Climate Change, for example, have the potential to provide only incremental improvements in policy guidance while imposing a heavy burden on the scientific community.

4. Metrics should serve to advance scientific progress or inquiry, not the reverse. Good measures will promote continuous improvements in the program, whereas poor measures could encourage actions to achieve high scores (e.g., "teaching to the test") and thereby lead to unintended consequences. For example, a metric to measure the convergence of climate models succeeds if it leads to an improved understanding of the physical processes being modeled, but fails if it subtly encourages researchers to adjust their models solely to bring them into better agreement with one another.

5. Metrics should be easily understood and broadly accepted by stakeholders. In standard land classifications, for instance, areas covered by dense canopy are considered "forest," even if they are severely logged and degraded. This land-cover metric would not be acceptable to paper companies, environmental groups, local governments, or other stakeholders without additional information on forest and environmental characteristics.

6. Promoting quality should be a key objective for any set of metrics. Quality is best assessed by independent, transparent peer review.

7. Metrics should assess process as well as progress. Metrics in a complex program such as the CCSP will be diverse, measuring factors that range from program planning, to resulting scientific knowledge and practical applications, to the ultimate impact of policy decisions on society.

8. A focus on a single measure of progress is often misguided. Relying solely on the metric of reducing uncertainty, for example, can create an erroneous sense of progress, since uncertainty can increase, decrease, or remain constant as the understanding of causal factors improves.

9. Considerable challenge should be expected in providing useful a priori outcome or impact metrics for discovery science. Care should be taken to avoid applying measures that stifle program elements for which the outcome is unknown. For example, metrics could have been devised to monitor the progress of C.D. Keeling's measurements of CO_2 in the atmosphere, two to four years after the program started. However, they would have missed the fundamental achievement enabled by this and subsequent measurements—the discovery of an annual cycle and decadal trend in atmospheric composition.

10. Metrics must evolve to keep pace with scientific progress and program objectives. Adjustments to the measures will be required as program managers gain experience and the program itself matures and evolves. For example, the CCSP strategic plan places greater emphasis on scientific assessments, decision support, and short-term outcomes than USGCRP plans and requires a greater breadth of metrics.

11. The development and application of meaningful metrics will require significant human, financial, and computational resources. It is possible to develop and apply thousands of metrics for the CCSP, but doing so would be costly and may not lead to improved program performance. A deliberative process of selecting the few most appropriate metrics, collecting the necessary information, and carrying out the evaluation will be required.

Although each of these principles is important, three merit especially careful attention: (1) leadership to guide the program and apply the metrics, (2) a plan of action against which to apply the measures, and (3) the potential to use metrics not just as simple measures of progress, but as tools to guide strategic planning and foster future progress. The first two are generally required if a program is to succeed. The last is a lesson learned from industry, and it informed the committee's approach to developing metrics for the CCSP.

METRICS FOR THE CCSP

The first challenge in developing metrics is to choose goals against which progress should be measured. The CCSP strategic plan has hundreds of goals and objectives, stated at different levels of specificity, from five overarching goals to 224 milestones, products, and payoffs. The committee found that the milestones, products, and payoffs could be grouped into eight themes, which cover the scope of the program and are amenable to the development of metrics:

1. improve data sets in space and time (e.g., create maps, databases, and data products; densify data networks);
2. improve estimates of physical quantities (e.g., through improvement of a measurement);
3. improve understanding of processes;
4. improve representation of processes (e.g., through modeling);
5. improve assessment of uncertainty, predictability, or predictive capabilities;
6. improve synthesis and assessment to inform;
7. improve assessment and management of risk; and
8. improve decision support for adaptive management and policy making.

One or two case studies were developed for each of these themes to explore how metrics could be developed and to determine how difficult it would be to generalize them to other elements of the program. The assumption was that a long list of unique metrics would emerge and that the challenge would be to choose and refine the few that seemed most important. A long list of metrics was in fact produced. However, comparison of the metrics revealed that many were similar (especially those that measured the research and development process or inputs to it) and that others could be rewritten more generically. This observation and subsequent tests led to a surprising conclusion: *a general set of metrics can be developed and used to measure progress and guide strategic thinking across the entire CCSP.* The general metrics recommended by the committee are given in Box ES.1.

Every metric will not be applicable to every CCSP program element. Moreover, it would be too expensive to measure and monitor all elements of the program, especially if the results are not going to be used. Consequently, efforts should be made to select the most appropriate measures. Metrics to guide strategic thinking will focus on identifying and monitoring program strengths and weaknesses with the object of enabling managers to make decisions that support successful outcomes. These measures will become apparent from even rough scores or answers to the metrics listed in Box ES.1. Metrics for demonstrating program progress will depend on how CCSP agencies define what constitutes success. As agencies gain experience, the initial metrics listed in Box ES.1 will be refined and simplified until only the most useful emerge.

Box ES.1
General Metrics for the CCSP

Process Metrics (measure a course of action taken to achieve a goal)

1. Leader with sufficient authority to allocate resources, direct research effort, and facilitate progress.
2. A multiyear plan that includes goals, focused statement of task, implementation, discovery, applications, and integration.
3. A functioning peer review process in place involving all appropriate stakeholders, with (a) underlying processes and timetables, (b) assessment of progress toward achieving program goals, and (c) an ability to revisit the plan in light of new advances.
4. A strategy for setting priorities and allocating resources among different elements of the program (including those that cross agencies) and advancing promising avenues of research and applications.
5. Procedures in place that enable or facilitate the use or understanding of the results by others (e.g., scientists in other disciplines, operational users, decision makers) and promote partnerships.

Input Metrics (measure tangible quantities put into a process to achieve a goal)

1. Sufficient intellectual and technologic foundation to support the research.
2. Sufficient commitment of resources (i.e., people, infrastructure, financial) directed specifically to allow the planned program to be carried out.
3. Sufficient resources to implement and sustain each of the following: (a) research enabling unanticipated scientific discovery, (b) investigation of competing ideas and interpretations, and (c) development of innovative and comprehensive approaches.
4. Sufficient resources to promote the development and maintenance of each of the following: (a) human capital; (b) measurement systems, predictive models, and synthesis and interpretive activities; (c) transition to operational activities where warranted; and (d) services that enable the use of data and information by relevant stakeholders.
5. The program takes advantage of existing resources (e.g., U.S. and foreign historical data records, infrastructure).

Output Metrics (measure the products and services delivered)

1. The program produces peer-reviewed and broadly accessible results, such as (a) data and information, (b) quantification of important phenomena or processes,

(c) new and applicable measurement techniques, (d) scenarios and decision support tools, and (e) well-described and demonstrated relationships aimed at improving understanding of processes or enabling forecasting and prediction.

2. An adequate community and/or infrastructure to support the program has been developed.

3. Appropriate stakeholders judge these results to be sufficient to address scientific questions and/or to inform management and policy decisions.

4. Synthesis and assessment products are created that incorporate these new developments.

5. Research results are communicated to an appropriate range of stakeholders.

Outcome Metrics (measure results that stem from use of the outputs and influence stakeholders outside the program)

1. The research has engendered significant new avenues of discovery.

2. The program has led to the identification of uncertainties, increased understanding of uncertainties, or reduced uncertainties that support decision making or facilitate the advance of other areas of science.

3. The program has yielded improved understanding, such as (a) more consistent and reliable predictions or forecasts, (b) increased confidence in our ability to simulate and predict climate change and variability, and (c) broadly accepted conclusions about key issues or relationships.

4. Research results have been transitioned to operational use.

5. Institutions and human capacity have been created that can better address a range of related problems and issues.

6. The measurements, analysis, and results are being used (a) to answer the high-priority climate questions that motivated them, (b) to address objectives outside the program plan, or (c) to support beneficial applications and decision making, such as forecasting, cost-benefit analysis, or improved assessment and management of risk.

Impact Metrics (measure the long-term societal, economic, or environmental consequences of an outcome)

1. The results of the program have informed policy and improved decision making.

2. The program has benefited society in terms of enhancing economic vitality, promoting environmental stewardship, protecting life and property, and reducing vulnerability to the impacts of climate change.

3. Public understanding of climate issues has increased.

The way in which general metrics are used depends on both the identity of the evaluators and the granularity of the program element being evaluated. Agency managers might give rough answers to all of the general metrics to assess strengths and weaknesses of the program and then determine an appropriate course of action. Indeed, the process of evaluating the program and selecting the measures should be as valuable to the agencies as the measures themselves. Expert panels might use the general metrics to develop a broader context for the project being reviewed. Finally, stakeholders might focus on outcome and impact metrics.

Highly focused programs may require highly specific metrics. The general metrics provide the categories to be evaluated, but they will have to be narrowed down and reworded in terms that are specific to the program goal. In refining the metrics, care must be taken to recognize and minimize biases, which are inevitable in subjective judgments. Attention must also be paid to developing an evaluation system to score each of the metrics and to aggregate different types of measures.

CONCLUSIONS

• Meaningful metrics can be developed for most aspects of the CCSP, from enhancement of data networks to increases in public awareness of climate change issues. The general set of metrics developed by the committee provides a useful starting point for identifying a small set of important measures, and the principles provide guidance for refining the metrics and avoiding unintended consequences.

• The metric used most commonly to gauge progress of the CCSP—reduction of uncertainty—has the potential to be misleading and should not be used in isolation. Uncertainty about future climate states may increase, decrease, or remain the same as more is understood about the elements that control the system.

• A mixture of qualitative and quantitative metrics is required to assess the progress of the CCSP. Quantitative measures (e.g., numerical scores, yes or no answers) are most useful for evaluating management, assessing the research and development process, or measuring aspects of research output. Qualitative measures are most useful for assessing quality and program results. In general, peer review is required to assess quality or progress toward improved understanding, and stakeholder judgments are required to assess the usefulness or impact of many programs.

• Discovery and innovation are difficult to measure with quantitative metrics. The best approach is to use process and input measures that ensure the promotion of discovery and innovation. As the science matures, more output, outcome, and impact measures become appropriate.

• A number of candidate CCSP metrics, especially those that assess outcomes and impacts, will depend on a wide range of factors, including some outside of the program (e.g., politics, technological advance). To avoid misinterpreting these measures (e.g., one weak component dominating the evaluation of an otherwise strong program), the explanation should accompany the score or answer. The context or explanation is as important as the score.

• Although some metrics can measure short-term impacts (e.g., CCSP payoffs scheduled to occur within two to fours years), it may take decades to fully assess the substantial contributions to the global debate on climate change being made by the CCSP and its predecessor USGCRP.

Although the maxim "what gets measured, gets managed" is not always true, the reverse generally is. A system of metrics, developed through an iterative process and evaluated in consultation with stakeholders, could be a valuable tool for managing the CCSP and for further increasing its usefulness to society.

1

Introduction

$\mathscr{T}$he Climate Change Science Program (CCSP) and its predecessor U.S. Global Change Research Program (USGCRP) have sponsored climate research and observations for nearly 15 years. Significant scientific discoveries and beneficial applications have resulted from these programs, but their overall progress has not been measured systematically. Metrics—simple qualitative or quantitative measures of performance with respect to a stated goal—offer a tool for gauging such progress, improving program performance, and demonstrating program successes to Congress, the Office of Management and Budget (OMB), and the public.

Metrics have long been used by industry to gauge the progress of research and development programs and to guide strategic planning. More recently, they have been used by universities to help make decisions on hiring and promoting faculty and by federal agencies to improve program performance and to increase public accountability. The latter was largely motivated by the Government Performance and Results Act (GPRA) of 1993, which required federal agencies to set strategic goals and to measure program performance against those goals.[1]

The GPRA does not apply to multiagency programs such as the CCSP or the USGCRP. However, the same motivating factors exist. CCSP agencies are striving (1) to demonstrate progress in climate change science, (2) to assess the current effectiveness of the program, and (3) to improve overall

---

[1]Public Law 103-62.

program performance.[2] Such an evaluation is needed to justify continued taxpayer support, especially in an era of declining budgets.

Studies in industry, academia, and the government suggest that metrics can be developed to document progress from past research programs and to evaluate future research performance.[3] The challenge is to create meaningful and effective metrics that accomplish the following:

- Convey an accurate view of scientific progress. A metric commonly used to evaluate advances in climate models, for example, is reduction of uncertainty of a projection or forecast.[4] However, progress in scientific and technical understanding can both increase and decrease uncertainty estimates.
- Result in a balance between high-risk research, long-term gain, and success in specific applications that are more easily measured.
- Accommodate the long time scales necessary for achieving results in basic research.

The following additional challenges are specific to the CCSP:

- To develop a methodology for creating metrics that can be applied to the entire CCSP. This is especially challenging because of the scope and diversity of the program. Thirteen agencies participate in the program, which encompasses a wide range of natural and social science disciplines, each of which has different approaches to and results from research, and activities ranging from observations, to basic research, to assessments and decision support (Box 1.1).
- To collect consistent data that can be used to assess and manage programs at the interagency level.

---

[2]Presentation to the committee by J. Mahoney, CCSP director, on December 17, 2003.

[3]Army Research Laboratory, 1996, Applying the Principles of the Government Performance and Results Act to the Research and Development Function: A Case Study Submitted to the Office of Management and Budget, 27 pp., <http://govinfo.library.unt.edu/npr/library/studies/casearla.pdf>; National Science and Technology Council, 1996, *Assessing Fundamental Science*, <http://www.nsf.gov/sbe/srs/ostp/assess/start.htm>; General Accounting Office, 1997, *Measuring Performance: Strengths and Limitations of Research Indicators*, GAO/RCED-97-91, Washington, D.C., 34 pp.; National Academy of Engineering and National Research Council, 1999, *Industrial Environmental Performance Metrics: Challenges and Opportunities*, National Academy Press, Washington, D.C., 252 pp.; National Research Council, 1999, *Evaluating Federal Research Programs: Research and the Government Performance and Results Act*, National Academy Press, Washington, D.C., 80 pp.; National Research Council, 2001, *Implementing the Government Performance and Results Act for Research: A Status Report*, National Academy Press, Washington, D.C., 190 pp.

[4]Presentations to the committee by J. Kaye, National Aeronautics and Space Administration, on December 17, 2003, and J. Rothenberg, OMB, on March 4, 2004.

**Box 1.1
CCSP Strategic Plan**

The CCSP strategic plan represents an attempt to integrate the science requirements of the USGCRP with the requirements laid out in the 2001 Climate Change Research Initiative to reduce uncertainty, improve observing systems, develop science-based resources to support policy making and resource management, and communicate findings to the broader community. The plan identifies five overarching goals that orient the research programs of 13 participating federal agencies around understanding climate change and managing its risks:

1. Improve knowledge of the Earth's past and present climate and environment, including its natural variability, and improve understanding of the causes of observed variability and change.
2. Improve quantification of the forces bringing about changes in the Earth's climate and related systems.
3. Reduce uncertainty in projections of how the Earth's climate and related systems may change in the future.
4. Understand the sensitivity and adaptability of different natural and managed ecosystems and human systems to climate and related global changes.
5. Explore the uses and identify the limits of evolving knowledge to manage risks and opportunities related to climate variability and change.

The plan also identifies four core approaches for working toward these goals:

1. Plan, sponsor, and conduct research on changes in climate and related systems.
2. Enhance observations and data management systems to generate a comprehensive set of variables needed for climate-related research.
3. Develop improved science-based resources to aid decision making.
4. Communicate results to domestic and international scientific and stakeholder communities, stressing openness and transparency.

Finally, the plan describes research needs in seven areas—atmospheric composition, climate variability and change, water cycle, land-use and land-cover change, carbon cycle, ecosystems, and human contributions and responses to environmental change—and specifies more than 200 milestones, products, and payoffs to be produced in these research areas within two to four years.

SOURCE: Climate Change Science Program and Subcommittee on Global Change Research, 2003, *Strategic Plan for the U.S. Climate Change Science Program*, Washington, D.C., 202 pp.

COMMITTEE CHARGE AND APPROACH

Given the challenges described above, how can progress in global change research be measured? At the request of James Mahoney, director of the Climate Change Science Program and chair of the Subcommittee on Global Change Research, the National Research Council convened an ad hoc committee to explore the issues and to recommend a methodology that agencies can use to demonstrate progress from past global change research investments and to institute meaningful and effective metrics for the future.[5] The committee was asked to avoid recommending changes to the CCSP strategic plan. The specific charge to the committee is given in Box 1.2.

The committee approached its charge first by examining what could be learned from previous efforts to develop metrics in federal government agencies, industry, and academia. Information was gathered from a literature review and briefings from agency program managers, climate change scientists, science historians, and policy experts. Based on this information, the committee identified principles for developing metrics for the CCSP. Special attention was given to issues such as peer review and reduction of uncertainty, which figure prominently in the metrics of each of these sectors as well as in the CCSP strategic plan.

Next, the committee chose case studies drawn from different parts of the CCSP strategic plan. The case studies ranged from collecting the data needed to better understand solar forcing of climate to improving adaptive management of water resources. For each case study, the committee developed example metrics and assessed the difficulty of applying them to other parts of the program. This exercise led to the development of a general set of metrics that could be used for the CCSP.

METRICS AND PERFORMANCE MEASURES

Metrics and performance measures gauge progress with respect to a stated goal. Therefore, they address the question: Is there demonstrable advancement in reaching a goal? Metrics and performance measures tend to be simple, focusing on a number, score, or a yes or no answer, but they can also integrate several different measures.[6]

Because the results of science and technology are both tangible and intangible, the associated metrics and performance measures may be quantitative or qualitative. The distinction between quantitative and qualitative

[5]Presentation to the committee by J. Mahoney, Climate Change Science Program, on December 17, 2003.
[6]Werner, B.M., and W.E. Souder, 1997, Measuring R&D performance—State of the art, *Research Technology Management*, **March-April**, 34–42.

Box 1.2
Committee Charge

Using the objectives of climate change and global change research as articulated in the CCSP strategic plan, the committee will develop quantitative metrics for documenting progress and evaluating future performance for selected areas of global change and climate change research. In particular, the study will

1. Provide a general assessment of how well CCSP objectives lend themselves to quantitative metrics.
2. Identify three to five areas of climate change and global change research that can and should be evaluated through quantitative performance measures.
3. For these areas, recommend specific metrics for documenting progress, measuring future performance (such as skill scores, correspondence across models, correspondence with observations), and communicating levels of performance.
4. Discuss possible limitations of quantitative performance measures for other areas of climate change and global change research.

In developing its recommendations, the committee will attempt to develop processes that can be applied in both the short term (e.g., two to four years) and longer terms, and will strive to avoid possible unintended consequences of performance measurement (e.g., unbalanced research portfolios, reduced innovation). The committee will not itself apply its proposed methodology to evaluate agency research efforts, although it may include in its report a few examples of how its recommended methods could be implemented.

---

is not always sharp, but in general, quantitative outputs (e.g., number of patents or new products) can be evaluated by direct measurement, whereas qualitative outputs (e.g., contributions to the pool of innovation, capabilities and skills of the scientific staff) require judgment to evaluate. Such judgments are subjective and lend themselves to scoring and, hence, some manipulation of quantities.[7]

In this report the term "metrics" is used for what some call "performance measures." As used by government agencies, performance measures include indicators and statistics that are used to assess progress toward pre-established goals. They tend to focus on "regularly collected data on the level and type of program activities, the direct products and services delivered

---

[7]Geisler, E., 2000, *The Metrics of Science and Technology*, Quorum Books, Westport, Conn., 380 pp.

TABLE 1.1 Example Definitions of Quantitative and Qualitative Metrics

Metric	Item Being Measured	Unit of Measurement	Inherent Value
Citation analysis	Scientific output	Citation counts	Impact of the work on the scientific community
Peer review	Scientific outcomes	Subjective analysis	Performance of scientists

by the program, and the results of those activities.[8] Because the results of scientific research are not easily defined in terms of performance and because a metric implies some ability to be quantitative, it seems a more apt terminology for use among scientists and by managers evaluating scientific programs. A metric is a "system of measurement that includes the item being measured, the unit of measurement, and the value of the unit."[9] Examples of the application of this definition to quantitative (citation analysis) and qualitative (peer review) metrics are given in Table 1.1.

Different types of metrics are used throughout industry, academia, and government. For example, OMB differentiates between long-term and annual measures and subdivides these categories into outcome and efficiency measures.[10] Academia relies on bibliometrics, which are published outputs such as number of journal articles or citations. This report focuses on five types of metrics—process, input, output, outcome, and impact—which are defined in Box 1.3.

ORGANIZATION OF THE REPORT

The purpose of this report is to provide a starting point for measuring progress of the CCSP and, by extension, its predecessor USGCRP. Chapter 2 describes different approaches that industry, academia, and federal agencies have taken to measure research performance. A more complete discussion of federal laws and policies driving government efforts to measure performance is given in Appendix A. Chapter 3 lays out principles for

---

[8]General Accounting Office, 2003, *Results-Oriented Government: Using GPRA to Address 21st Century Challenges*, GAO-03-1166T, Washington, D.C., p. 9.

[9]Geisler, E., 2000, *The Metrics of Science and Technology*, Quorum Books, Westport, Conn., pp. 74–75.

[10]Process and output measures are also allowed in some cases. See Office of Management and Budget, 2005, Guidance for Completing the Program Assessment Rating Tool (PART), pp. 9–10, <http://www.whitehouse.gov/omb/ part/fy2005/2005_guidance.doc>.

Box 1.3
Categories of Metrics Used in This Report

Metrics can be devised to evaluate the overall process for reaching a goal, or any stage or result of the process (input, output, outcome, impact). Definitions of these categories and example metrics related to the discovery of the Antarctic ozone hole are given below.

1. **Process**—a course of action taken to achieve a goal. Example metrics include existence of a project champion and length of time between starting the research and delivering an assessment on stratospheric ozone depletion to policy makers.

2. **Input**—tangible quantities put into a process to achieve a goal. An example input metric is expenditures for (a) theoretical and laboratory studies on ozone production and destruction, (b) development and deployment of sensors to sample the stratosphere, (c) modeling and analysis of data, or (d) meetings and publications.

3. **Output**—products and services delivered. Examples of output metrics include number of models that take into account new findings on chlorofluorocarbon chemistry or number of publications and news reports on the cause of stratospheric ozone depletion and its possible consequences.

4. **Outcome**—results that stem from use of the outputs. Unlike output measures, outcomes refer to an event or condition that is external to the program and is of direct importance to the intended beneficiaries (e.g., scientists, agency managers, policy makers, other stakeholders). Examples of outcome metrics are the number of alternative refrigerants introduced to society to reduce the loss of stratospheric ozone and scientific outputs integrated into a new understanding of the causes of the Antarctic ozone hole.

5. **Impact**—the effect that an outcome has on something else. Impact metrics are outcomes that focus on long-term societal, economic, or environmental consequences. Examples of impact metrics include the recovery of stratospheric ozone resulting from implementation of the Montreal Protocol and related policies and the increase in public understanding of the causes and consequences of ozone loss.

---

developing metrics, based on the experience of industry, academia, and federal agencies. Chapter 4 focuses on the metric most commonly used to measure progress in climate science: uncertainty reduction. Chapter 5 describes the process by which the committee developed metrics and summarizes conclusions from developing metrics in case studies that appear here and in Appendix B. A set of general metrics for assessing the progress of CCSP program elements and for guiding future strategic planning is proposed and tested in Chapter 6. Additional metrics developed elsewhere for science and technology programs in general are presented in Appendix C. Finally, Chapter 7 presents answers to the questions in the committee's charge and discusses implementation issues.

2

Lessons Learned from Developing Metrics

*I*ndustry, academia, and federal agencies all have experience in measuring and monitoring research performance. This chapter describes lessons learned from these sectors as well as insight from retrospective analysis of the stratospheric ozone program of the 1970s and 1980s that might be useful to the Climate Change Science Program (CCSP).

INDUSTRY RESEARCH

Use of Metrics in Manufacturing

For more than 200 years,[1] industry has employed metrics to monitor budget, safety, health, environmental impacts, material, energy, and product quality.[2] A study group of 13 companies has been meeting since 1998 to

---

[1]DuPont, E.I., 1811, Workers' rules, Accession 146, Hagley Museum, Manuscripts and Archives Division, Wilmington, Del.; Hounshell, D.A., and J.K. Smith, 1988, Science and Corporate Strategy, DuPont, p. 2.; Kinnane, A., 2002, *DuPont: From the Banks of the Brandywine to Miracles of Science*, E.I. DuPont, Wilmington, Del., 268 pp.

[2]Examples of financial metrics can be found in the annual report of almost any major chemical company. Quality management metrics appear in the International Organization for Standardization's ISO 9000 (<http://www.iso.ch/iso/en/iso9000-14000/index.html>). Examples of safety, health, environmental, material consumption, and energy consumption metrics are given in National Academy of Engineering (NAE) and National Research Council, 1999, *Industrial Environmental Performance Metrics: Challenges and Opportunities*, National Academy Press, Washington, D.C., 252 pp.

identify metrics that could be useful tools for industry.[3] The group found that the development of useful metrics in the manufacturing sector begins with careful formulation of the objectives for creating them. Important questions to be considered include the following:

- What is the purpose of the measurement effort?
- What are the "issues" to be measured?
- How are goals set for each issue?
- How is performance measured for that issue?
- How should the metric be compared to a performance standard?
- How will the metric be communicated to the intended audience?

Metrics that have proven useful in the manufacturing sector tend to have the following attributes:

- few in number, to avoid confusing the audience with excessive data;
- simple and thus easily understood by a broad audience;
- sufficiently accurate to be credible;
- an agreed-upon definition;
- relatively easy to develop, preferably using existing data;
- robust and thus requiring minimal exceptions and footnotes; and
- sufficiently durable to remain relatively constant over the years.

Metrics used in manufacturing tend to focus on input, output, or process (see definitions in Box 1.3), and they are commonly normalized to enable comparisons. In general, output metrics (e.g., pounds of product per pound of raw material purchased) have been the most successful because they are highly specific, relatively unambiguous, and directly related to a specific end point. Over time, and frequently after adjustment based on learning, the use of metrics in the manufacturing sector has been so effective as to give rise to the maxim "what gets measured, gets managed."

Extension to Research and Development

Success in the manufacturing sector encouraged efforts to develop quantifiable metrics for research and development (R&D) beginning in the late 1970s.[4] However, problems immediately arose. The most successful manu-

---

[3]The group meets under the sponsorship of the American Institute of Chemical Engineers' (AIChE) Center for Waste Reduction Technology. See reports on AIChE collaborative projects, focus area on sustainable development at <http://www.aiche.org/cwrt/pdf/BaselineMetrics.pdf>.

[4]Blaustein, M.A., 2003, Managing a breakthrough research portfolio for commercial success, Presentation to the American Chemical Society, March 25, 2003; Miller, J., and J. Hillenbrand,

facturing metrics measured a discrete item of output that could be produced in a short amount of time. These conditions are difficult to achieve in R&D. Research outputs are far less easily defined and quantified than manufacturing outputs, and the proof that a particular metric measured something useful, such as a profitable product or an efficient process, might take years.

Early metrics proposed for R&D included the following:

- *Input metrics*: total expenses or other resources employed, expenses or resources consumed per principal investigator (PI), and PI activities such as number of technical meetings attended.
- *Output metrics*: number of compounds or materials made or screened, and number of publications or patents per PI.
- *Outcome metrics*: PI professional recognition earned.

The list of possible metrics was long and failures were common. For example, "number of compounds made" could lead to an emphasis on "easy" chemistry instead of groundbreaking effort in a difficult, but potentially fruitful, area. Moreover, absent any professional judgment on the relevance and quality of items such as "technical meetings," measurement of these items merely consumed time and money that might have been better spent elsewhere.

Based on these early lessons learned, a small number of process and output metrics emerged that proved useful to some businesses. These included

- elapsed time to produce and register a quality product, from discovery to commercialization; and
- creation of an idealized vision for processing operations such as no downtime, no in-process delays, or zero emissions. Although such goals, stated as process metrics, might not be reachable, they serve to drive research in a desirable direction.

Ultimately, most R&D metrics fell from favor because of the long period between measurement and analysis of the result of R&D and the need for expert judgment in evaluating the quality of the item being measured. They are being widely replaced by a "stage-gate" approach for managing R&D. In the stage-gate approach, the R&D process is divided into three or more stages, ranging from discovery through commercialization (see Table 2.1). The number of stages is usually specific to the business, with the number increasing as the complexity and length of the R&D process increase. For example, there will be more stages in the R&D process

2000, Meaningful R&D output metrics: An unmet need of technology and business leadership, Presentation at the Corporate Technology Council, E.I. DuPont, June 20, 2000.

TABLE 2.1 Example of the Stage-Gate Steps and Metrics for R&D in a Traditional Advanced Materials Chemical Industry

Metric Theme	Stage 1 Feasibility	Stage 2 Confirmation	Stage 3 Commercialization
Sustainable product	• Customer needs have been analyzed • Improved properties, identified through analysis of customer needs, such as increased strength and corrosion or stain resistance, have been demonstrated	• New discovery has led to an established patent position • Manufacturing or marketing strengths of the new discovery have been analyzed	• Customer alternatives to the use of the new product have been analyzed
Economics	• Product or process concept has been proven, even though economic practicality has not been established	• Materials cost, process yield, catalyst life, capital intensity, and competitors have been analyzed	• Sustained pilot operation has been achieved • Impurities and recycle streams have been analyzed
Customer acceptance	• Target customers have been identified	• Plan exists for partnerships and for access to market • Customer reaction to prototype has been satisfactory	• Partnerships and access to market have been established
Safety and environment	• Alternative materials have been considered • Radical processing concept has been considered • Safety in use has been analyzed	• Inherently safe and "green" concepts have been demonstrated • Toxicology tests have been completed	• Design exists for "fail-safe" operation and "zero" emissions

of a drug company than in the R&D process of a polymer developing company. Groups of metrics are identified within each stage, and a satisfactory response to the metrics must be achieved before the project is allowed to proceed to the next stage. Advancement to successive stages can easily be tracked and converted to process metrics reflecting the status or progress of a program (e.g., a yes or no answer to whether the program has been completed). The main difficulty with the stage-gate approach is in choosing the metric themes for each stage and assessing the quality of the results, both of which require professional judgment.

A stage-gate process such as that illustrated in Table 2.1 is generally initiated by the scientists in the organization, following an R&D discovery or a promising analysis.[5] Generally, after a year of effort by a single principal investigator, the program either transitions to a managed stage-gate R&D program or is terminated due to apparent infeasibility or poor fit with the business intentions of the company. Whether or not an R&D project advances to Stage 1 depends on demonstration of the following:

- technical feasibility,
- scientific uniqueness,
- availability of skills within the organization required to bring the research to fruition,
- ability to identify a market within the growth areas promoted by the company,
- ability to define realistic goals and objectives and to establish a clear focus and targets, and
- ability to attract sponsorship by the business unit of the organization.

Tools for Strategic Analysis

Industry commonly uses metrics to guide strategic planning. Lessons learned from this experience include the following:

- Metrics can be applied to most ongoing operations. The greatest value of metrics will be achieved by selecting a few key issues and monitoring them over a long time.
- Data for measuring research progress are generally of poor quality initially. Standardization of data collection, quality assessment, and verification are necessary to produce broadly credible results.

---

[5]Blaustein, M.A., 2003, Managing a breakthrough research portfolio for commercial success, Presentation to the American Chemical Society, March 25, 2003; Carberry, J., 2004, Managing research programs via numerical metrics and/or a "stage gate" process," Presentation to the NAE Committee on Global Climate Change R&D, March 3, 2004.

•   Most successful R&D programs measure progress against a clearly constructed business plan that includes a statement of the task, goals and milestones, budget, internal or external peer review plan, and communication plan.

Applicability to the CCSP

The industrial experience with metrics has much to offer the CCSP. For example, the attributes of manufacturing metrics (e.g., few metrics, easily understood) and the importance of expert judgment in assessing the relevance and quality of process and output metrics are likely to be widely applicable. In addition, a number of industry approaches (e.g., analysis of program resource distribution, use of R&D process metrics and peer review rankings, graphical summaries) could be used to guide strategic planning and improve R&D quality and progress. Finally, a stage-gate process might be used to help CCSP agencies plan how to move a program emphasis from the discovery phase that precedes Stage 1 (feasibility of using basic research results to improve decision making), to Stage 2 (developing and testing decision-making tools), to Stage 3 (decision making and communicating program results).

Following is a hypothetical example related to the CCSP. Suppose that R&D funding is $900 million and that it is divided among the CCSP goals and approaches according to Table 2.2 (shown graphically Figure 2.1).

TABLE 2.2 Hypothetical Distribution of Funding Applied to CCSP Overarching Goals and Core Approaches

CCSP Goals→	Funding (millions of dollars)					
Approaches↓	Improve Knowledge	Improve Quantification	Reduce Uncertainty	Understand Adaptability	Manage Risk	Percentage
Fundamental research	100	85	20	40	37	31
Enhance observations	72	63	73	24	45	31
Aid decision making	19	37	48	65	80	28
Communicate results	26	17	16	14	19	10
Percentage	25	22	17	16	20	

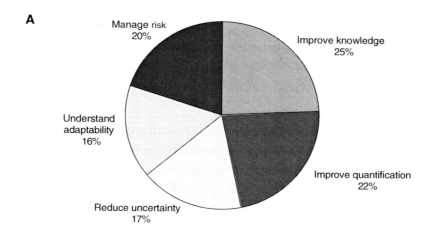

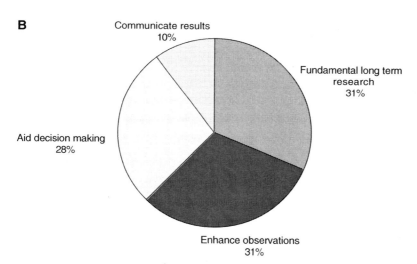

FIGURE 2.1 Distribution of effort (A) in the five CCSP overarching goals and (B) in the four CCSP core approaches, based on hypothetical data in Table 2.2.

From these data, program managers would decide if the distribution of effort is appropriate or if adjustments are needed. They may decide, for example, that too little of the effort is focused on communicating results.

Once program managers are satisfied, the process of evaluating the quality of research activities can begin. Again for the hypothetical example above, assume that the $63 million to improve quantification or enhance observations (Table 2.2) is divided among six research projects. Assume also that 11 R&D measures of the management and leadership process have been developed and scored by peer review as shown in Table 2.3. The peer review panel might evaluate all six projects and rate the quality of the project management on, for example, a 1 to 5 scoring system. An illustration of that scoring system follows:

R&D Metric: Quality of the Internal or External Review Process for This Task
   1 = Poor—no review plan in place; no reviews, even ad hoc
   2 = Fair—no review plan in place; infrequent, ad hoc reviews; unreliable follow-up
   3 = Average—review plan exists; irregularly followed; unreliable follow-up
   4 = Good—plan exists; regularly followed; spotty follow-up
   5 = Excellent—plan exists; regularly followed; excellent follow-up

TABLE 2.3 Hypothetical Example of 11 R&D Process Metrics Applied to Six Research Projects

Metric	Project 1	Project 2	Project 3	Project 4	Project 5	Project 6	Average
Quality of the *internal or external peer review* process for this task	4	5	3	2	2	4	3.3
Statement of the task is sufficiently focused and specific to be evaluated by the peer review process	4	5	3	3	3	3	3.5
Quality of the selection and definition of *long-term goals*	4	5	2	1	3	2	2.8
Quality of the selection and definition of *milestones*	3	3	2	1	1	2	2.0

TABLE 2.3 Continued

Metric	Project 1	Project 2	Project 3	Project 4	Project 5	Project 6	Average
Progress in achieving milestones	3	3	1	1	1	2	1.8
Communication of the work	1	4	1	2	1	3	2.0
Projected cost to completion in relation to relative importance of the subject and total funds that might be available	2	4	1	3	3	3	2.7
Usefulness of the results in meeting the overall goal	4	5	1	2	3	3	3.0
Feasibility of completing the work in a time frame useful for the overall study	4	4	2	2	4	3	3.2
What is the assessment of the *scientific quality* of the work?	4	5	3	4	2	3	3.5
What is the assessment of the *performance versus the technical specification*?	4	4	2	3	3	3	3.2
Average	3.5	4.5	2.2	2.5	2.8	3.4	2.8

NOTE: Rankings are given on a 1-5 (poor to excellent) scale.

From the information in Table 2.3 program managers could begin asking critical questions about the quality of the R&D effort, for example:

- Is research project 3 so weak based on the average score that it should be discontinued or at least supervised more closely?
- Why are the scores for progress in achieving milestones (fifth measure) uniformly low and what can be done?

UNIVERSITY RESEARCH

Metrics in academia are used to assess the performance of faculty, departments, and the university itself, as well as to manage resources. Metrics to evaluate the success of a university generally focus on outcomes and impacts, such as fraction of degrees completed, student satisfaction, success of the graduates, and national reputation.[6]

Faculty appointment and promotion systems are designed to evaluate a number of activities, including research, teaching, and service. Teaching and service metrics generally focus on outputs (e.g., number of undergraduates taught, courses developed, or committees served on), although judgment is required to assess the quality of teaching and to weigh the prestige of teaching awards and committee memberships. Peer review is the foundation of research assessment (Box 2.1), and it usually takes the form of internal committees that both review the person's work and take account of outside letters of evaluation from experts in fields relevant to the particular candidate. These evaluations require a good deal of personal judgment— about qualities of mind, the influence of particular ideas or writings, and the person's promise for future contributions—but usually these subjective judgments are bolstered by metrics of research performance. Examples of research metrics include the following:

- number of articles or books that have been accepted in the published literature;
- the subset of articles that have appeared in the "top" journals in a field (i.e., those viewed as having the toughest review);

[6]A number of reports rank universities by reputational measures such as the quality of research programs (e.g., National Research Council, 1995, *Research—Doctorate Programs in the United States: Continuity and Change*, National Academy Press, Washington, D.C., 768 pp.) or other characteristics, such as selection, retention, and graduation of students; faculty resources; and alumni giving (e.g., U.S. News and World Report, 2005, Best Colleges Index, <http://www.usnews.com/usnews/edu/college/rankings/rankindex_brief.php>). A major criticism of such national rankings is that they distract universities from trying to improve scholarship. See National Research Council, 2003, *Assessing Research—Doctorate Programs: A Methodology Study*, National Academies Press, Washington, D.C., 164 pp.

Box 2.1
Scholarly Peer Review

Peer review is generally defined as a critical evaluation by independent experts of "the technical merit of research proposals, projects, and programs."[a] A mainstay of the scientific process, peer review provides an "in-depth critique of assumptions, calculations, extrapolations, alternate interpretations, methodology and acceptance of criteria employed and conclusions drawn in the original work."[b] While the focus on scientific expertise is paramount, commentators also note that the "peer review process is invariably judgmental and thus inevitably involves interplay between expert and personal judgments."[c]

Definitions of peer review generally focus on the independence and the appropriate expertise of the peer reviewer. An adequate peer review satisfies three criteria: (1) it includes multiple assessments, (2) it is conducted by scientists who have expertise in the research in question, and (3) the scientists conducting the review have no direct connection to the research or its sponsors.[c]

The second criterion can be difficult to fulfill in evaluations of interdisciplinary work. Even if a peer review group with all of the relevant disciplines is assembled, its members may have difficulty seeing beyond the boundaries of their own disciplines to properly evaluate the integrated product. Ideally, each member of the evaluation group would invest significant time developing at least a basic understanding of the other relevant fields. However, this is a luxury that peer review committees rarely, if ever, have. The ideal of unconflicted peer review (criterion 3) is also usually not achieved, simply because there is a limited pool of experts and those most knowledgeable are also likely to be connected to the research and its sponsors. In such cases the objective becomes one of minimizing conflict of interest and bias.

---

[a]National Research Council, 1998, *Peer Review in Environmental Technology Development Programs*, National Academy Press, Washington, D.C., p. 2.
[b]Altman, W.D., J.P. Donnelly, and J.E. Kennedy, 1988, Peer Review for High-Level Nuclear Waste Repositories: Generic Technical Position, Nuclear Regulatory Commission, NUREG-1297, Washington, D.C., p. 2.
[c]Salter, L., 1985, Science and peer review: The Canadian standard-setting experience, *Science, Technology and Human Values*, **10**, 37–46.

- number of other publications—including book chapters, conference proceedings, and research reports—that may not have been subjected to peer review;
- number of citations;
- number of honors and awards; and
- amount of extramural funding.[7]

---

[7]National Research Council, 1995, *Research—Doctorate Programs in the United States: Continuity and Change*, National Academy Press, Washington, D.C., 768 pp.; Graham, H.D., and N. Diamond, 1997, *The Rise of American Research Universities: Elites and Challenges in the Postwar Era*, Johns Hopkins University Press, Baltimore, Md., 319 pp.

By normalizing these metrics to the number of faculty members, it is possible to measure departmental performance and productivity. These metrics are used to make resource decisions, such as how much space to allocate to different departments, and to compare programs at different institutions.

Such metrics are useful but have significant shortcomings. For example, judgments about the quality and impact of publications are difficult because of the long lags in the publication process and even longer lags in scientific appreciation of the work. Moreover, results are influenced by the membership of the peer evaluation group and the process by which members are selected. This is especially true of assessments of interdisciplinary work (see Box 2.1). Nevertheless, peer review committees, supported by output and impact metrics, have proven to be the most successful means of evaluating research progress in most academic disciplines.[8]

Applicability to the CCSP

Peer review, supplemented by output and impact metrics, has generally been successful in evaluating academic research progress. Its most effective use is in assessing research plans, the potential for proposed research to succeed, and the quality of research results. Expert opinion is also essential for assessing long-term term research outcomes or impact or for determining when changes in research direction are required. This approach should apply equally well to CCSP basic science programs. Many, perhaps most, individual studies will require only the customary peer review of each discipline. However, expert groups with both the right expertise and sufficient time to learn about other relevant fields will be required to make informed judgments about cross-cutting problems such as climate change. Assessments of the human and environmental consequences of climate change, for example, will require experts in risk assessment, as well as in geophysics, biology, chemistry, socioeconomics, and statistics.

FEDERAL AGENCY RESEARCH

Measuring Government Performance

Some federal agencies have been collecting data to characterize and evaluate their scientific activities for decades.[9] However, it was only with

---

[8]Bozeman, B., 1993, Peer review and evaluation of R&D impacts, in *Evaluating R&D Impacts: Methods and Practice*, B. Bozeman and J. Melkers, eds., Kluwer Academic Publishers, Boston, Mass., pp. 79–98.

[9]Examples include R&D strategic planning and program review carried out by the Depart-

the advent of the Government Performance and Results Act (GPRA) of 1993 and related Office of Management and Budget (OMB) policies that a concerted effort was made to measure progress throughout the federal government (Appendix A).

Federal agencies evaluate research programs to improve management and demonstrate to bureaucratic superiors and Congress that their programs have produced benefits that justify their cost. The preferred form of evaluation is peer review (Boxes 2.1 and 2.2), and OMB has stipulated that peer review of research and development programs be of high quality, of sufficient scope, unbiased and independent, and conducted on a regular basis.[10] Performance measures include both qualitative (e.g., productivity, research quality, relevance of research to the agency's mission, leadership)[11] and quantitative measures.[12] However, quantitative measures have proven to be difficult to apply on an annual basis for the following reasons:

- The discovery and innovation process is complex and often involves many factors that are not related to research (e.g., marketing, intellectual property rights).
- It may take many years for a research project to achieve results.
- Outcomes are not always directly traceable to specific inputs or

---

ment of Defense since the 1960s and later by the National Science Foundation and the National Institutes of Health. See Cetron, M.J., J. Martino, and L. Roepcke, 1967, The selection of R&D program content—Survey of quantitative methods, *IEEE Transactions on Engineering Management*, EM-14, 4–13; Office of Technology Assessment, 1986, *Research Funding as an Investment: Can We Measure the Returns?* OTA-TMSET, U.S. Congress, Washington, D.C., 72 pp.; Kostoff, R., 1993, Evaluating federal R&D in the United States, in *Evaluating R&D Impacts: Methods and Practice*, B. Bozeman and J. Melkers, eds., Kluwer Academic Publishers, Boston, Mass., pp. 163–178; National Science Foundation, 2004, Science and Engineering Indicators 2004, NSB 04-07, <http://www.nsf.gov/sbe/srs/seind04/start.htm>.

[10]Office of Management and Budget, 2005, Guidance for Completing the Program Assessment Rating Tool (PART), pp. 28–30, <http://www.whitehouse.gov/omb/part/fy2005/2005_guidance.doc>.

[11]Army Research Laboratory, 1996, Applying the Principles of the Government Performance and Results Act to the Research and Development Function: A Case Study Submitted to the Office of Management and Budget, 27 pp., <http://govinfo.library.unt.edu/npr/library/studies/casearla.pdf>; National Research Council, 1999, *Evaluating Federal Research Programs: Research and the Government Performance and Results Act*, National Academy Press, Washington, D.C., 80 pp.; Memorandum on FY 2004 interagency research and development priorities, from John H. Marburger III, director of the Office of Science and Technology Policy, and Mitchell Daniels, director of the Office of Management and Budget, on May 30, 2002, <http://www.ostp.gov/html/ombguidmemo.pdf>.

[12]Roessner, D., 1993, Use of quantitative methods to support research decisions in business and government, in *Evaluating R&D Impacts: Methods and Practice*, B. Bozeman and J. Melkers, eds., Kluwer Academic Publishers, Boston, Mass., pp. 179–205.

Box 2.2
Agency Review Processes

Federal agencies employ several kinds of review. In addition to peer review,[a] agencies use the following methods to gain information and feedback:

*   *Internal reviews*—these reviews are conducted by scientific or technical experts and managers within the agency and are used to manage programs.
*   *Notice and comment*—proposed regulations, including any underlying scientific analyses that have shaped the proposed regulation, are published in the *Federal Register* for public comment.[b] Comments are submitted in a variety of forms, including published scientific articles, "gray" literature, letters from Nobel laureates, and letters from the general public. These submissions are a valuable source of information and opinion, but not all have been peer reviewed or written by independent individuals with expertise in the relevant research.
*   *Stakeholder processes*—informal processes, such as "town hall" meetings, are a useful source of information and opinion on regulation development.[c] Like notice-and-comment, these processes invite input from any source, independent or not, expert or not, and external or not. Scientists often participate in these processes, sometimes as paid consultants and sometimes as individuals expressing politically oriented views of scientific analyses and policies.

The kind(s) of review used by an agency depends on circumstances. For example, defense agencies (and industry) rely heavily on internal reviews because of the highly specialized and/or confidential nature of the research. Regulatory agencies use notice-and-comment and stakeholder processes because public input from a wide variety of sources is desirable and/or mandated by law. In some cases, the type of review chosen depends on timing or funding constraints. Tight deadlines (court-ordered, administrative, political) can deter peer review planning and shorten or abort scheduled peer reviews. Also, insufficient funding may tempt agencies to substitute less costly review processes for peer review.

---

[a] General Accounting Office, 1999, *Federal Research: Peer Review Practices at Federal Science Agencies*, GAO/RCED-99-99, Washington, D.C., 71 pp. An example of agency peer review guidelines is given in Environmental Protection Agency, 2000, *Science Policy Council Handbook: Peer Review*, 2nd ed., EPA 100-B-00-001, National Service Center for Environmental Publications, Cincinnati, OH, 188 pp., <www.EPA.gov>. Guidelines that are applicable to all federal agencies are given in Office of Management and Budget, 2004, Revised Information Quality Bulletin for Peer Review, 36 pp., <http://www.whitehouse.gov/omb/inforeg/peer_review041404.pdf>.
[b] The requirement for public participation in the development of federal regulations is laid out in the Administrative Procedures Act, 5 USC 553.
[c] Till, J.E., 1995, Building credibility in public studies, *American Scientist*, **83**, 468–473.

may result from a combination of inputs. As a result, the results of research are not usually predictable.

• Negative findings or research results that contribute to objectives in other parts of the government are not valued as outcomes in the GPRA sense.

• Information needed for an assessment may be unobtainable, inconsistent, or ambiguous.[13]

Federal agencies are responsible for developing performance measures both for their annual performance plans (a requirement of GPRA) and for OMB's Program Assessment Rating Tool (PART) (see Appendix A). GPRA performance measures are relatively numerous and broad in scope (e.g., they may include process, output, and outcome measures). Approaches to GPRA measures vary, but most federal agencies are using expert review to judge progress in research and education, and are developing quantitative measures to evaluate management and process (e.g., increase award size or duration).[14] An example of different approaches to GPRA performance measures used by the National Science Foundation (NSF) and the National Oceanic and Atmospheric Administration (NOAA) is given in Table 2.4.

The PART measures provide OMB with a uniform approach to assessing and rating programs across the federal government. They are consistent with GPRA measures, but are few in number, reflect program priorities, and focus on outcomes.[15] If outcome measures cannot be devised, OMB allows research and development agencies to substitute output or process measures (e.g., Questions 2.3, 3.1, and 3.2 in Box A.2, Appendix A). The fiscal year (FY) 2005 PART measures associated with the climate change programs of participating agencies are given in Table 2.5.

The use of performance measures in the government has had mixed success. A 2004 General Accounting Office report found that R&D man-

---

[13]Army Research Laboratory, 1996, Applying the Principles of the Government Performance and Results Act to the Research and Development Function: A Case Study Submitted to the Office of Management and Budget, 27 pp., <http://govinfo.library.unt.edu/npr/library/studies/casearla.pdf>; National Science and Technology Council, 1996, Assessing Fundamental Science, <http://www.nsf.gov/sbe/srs/ostp/assess/start.htm>; General Accounting Office, 1997, *Measuring Performance: Strengths and Limitations of Research Indicators*, GAO/RCED-97-91, Washington, D.C., 34 pp.; National Research Council, 1999, *Evaluating Federal Research Programs: Research and the Government Performance and Results Act*, National Academy Press, Washington, D.C., 80 pp.; National Research Council, 2001, *Implementing the Government Performance and Results Act for Research: A Status Report*, National Academy Press, Washington, D.C., 190 pp.

[14]Presentations to the committee by S. Cozzens, Georgia Institute of Technology, C. Robinson, National Science Foundation, and C. Oros, U.S. Department of Agriculture's Cooperative State Research, Education, and Extension Service, on March 4, 2004.

[15]See <http://www.whitehouse.gov/omb/part/fy2005/2005_guidance.doc>.

TABLE 2.4 Contrasting Types of Government Performance and Results Act Measures at NSF and NOAA

Strategic Goal	Annual Performance Goal	Performance Measure
National Science Foundation[a]		
People: A diverse, competitive, and globally engaged U.S. work force of scientists, engineers, technologists, and well-prepared citizens	NSF will demonstrate significant achievement for the majority of the following performance indicators related to the people outcome goal	• Promote greater diversity in the science and engineering (S&E) work force through increased participation of underrepresented groups in NSF activities • Support programs that attract and prepare U.S. students to be highly qualified members of the global S&E work force, including providing opportunities for international study, collaborations, and partnerships • Promote public understanding and appreciation of science, technology, engineering, and mathematics and build bridges between formal and informal science education • Support innovative research on learning, teaching, and education that provides a scientific basis for improving science, technology, engineering, and mathematics education at all levels • Develop the nation's capability to provide K-12 and higher-education faculty with opportunities for continuous learning and career development in science, technology, engineering, and mathematics
Ideas: Discovery across the frontier of science and engineering, connected to learning, innovation, and service to society	NSF will demonstrate significant achievement for the majority of the following performance indicators related to the ideas outcome goal	• Enable people who work at the forefront of discovery to make important and significant contributions to science and engineering knowledge • Encourage collaborative research and education efforts across organizations, disciplines, sectors, and international boundaries • Foster connections between discoveries and their use in the service of society • Increase opportunities for individuals from underrepresented groups and institutions to conduct high-quality, competitive research and education activities • Provide leadership in identifying and developing new research and education opportunities within and across science and engineering fields • Accelerate progress in selected science and engineering areas of high priority by creating new integrative and cross-disciplinary knowledge and tools and by providing people with new skills and perspectives

Tools: Broadly accessible, state-of-the-art S&E facilities, tools, and other infrastructure that enable discovery, learning, and innovation	NSF will demonstrate significant achievement for the majority of the following performance indicators related to the tools outcome goal	• Expand opportunities for U.S. researchers, educators, and students at all levels to access state-of-the art S&E facilities, tools, databases, and other infrastructure • Provide leadership in the development, construction, and operation of major next-generation facilities and other large research and education platforms • Develop and deploy an advanced cyberinfrastructure to enable all fields of science and engineering to fully utilize state-of-the-art computation • Provide for the collection and analysis of the scientific and technical resources of the United States and other nations to inform policy formulation and resource allocation • Support research that advances instrument technology and leads to the development of next-generation research and education tools

National Oceanic and Atmospheric Administration[b]

Protect, restore, and manage the use of coastal and ocean resources through ecosystem-based management	Improve protection, restoration, and management of coastal and ocean resources through ecosystem-based management	• Number of overfished major stocks of fish reduced to 42 • Number of major stocks with an "unknown" stock status reduced to 77 • Percentage of plans to rebuild overfished major stocks to sustainable levels increased to 98% • Number of threatened species with lowered risk of extinction increased to 6 • Number of commercial fisheries that have insignificant marine mammal mortality maintained at 8 • Number of endangered species with lowered risk of extinction increased to 7 • Number of habitat acres restored (annual/cumulative) increased to 4,500/19,280
Understand climate variability and change to enhance society's ability to plan and respond	Increase understanding of climate variability and change	• U.S. temperature forecasts (cumulative skill score computed over the regions where predictions are made) improved to 22 • New climate observations introduced increased to 355 • Reduce uncertainty of atmospheric estimates of U.S. carbon source/sink to ±0.5 gigatons carbon per year • Improve measurements of North Atlantic and North Pacific Ocean basin CO_2 fluxes to within ±0.1 petagram carbon per year • Capture more than 90% of true contiguous U.S. temperature trend and more than 70% of true contiguous U.S. precipitation

continued

TABLE 2.4 Continued

Strategic Goal	Annual Performance Goal	Performance Measure
Serve society's needs for weather and water information	Improve accuracy and timeliness of weather and water information	• Lead time (increased to 13 minutes), accuracy (increased to 73%), and false alarm rate (decreased to 69%) for severe weather warnings for tornadoes • Increased lead time (53 minutes) and accuracy (89%) for severe weather warnings for flash floods • Hurricane forecast track error (48 hour) reduced to 128 • Accuracy (threat score) of day 1 precipitation forecasts increased to 27% • Increased lead time (15 hours) and accuracy (90%) for winter storm warnings • Cumulative percentage of U.S. shoreline and inland areas that have improved ability to reduce coastal hazard impacts increased to 28%

aNational Science Foundation FY 2005 Budget Request to Congress, <http://www.nsf.gov/bfa/bud/fy2005/pdf/fy2005.pdf>.
bNational Oceanic and Atmospheric Administration FY 2005 Annual Performance Plan, <http://www.osec.doc.gov/bmi/budget/05APP/NOAA05APP.pdf>.

TABLE 2.5 Climate Science-Related Performance Measures in OMB's FY 2005 PART

Agency	Performance Measure[a]
DOE	• Progress in delivering improved climate data and models for policy makers to determine safe levels of greenhouse gases and, by 2013, toward substantially reducing differences between observed temperature and model simulations at subcontinental scales using several decades of recent data. An independent expert panel will conduct a review and rate progress (excellent, adequate, poor) on a triennial basis
EPA	• Million metric tons of carbon equivalent of greenhouse gas emissions reduced in the building (or industry or transportation) sector • Tons of greenhouse gas emissions prevented per societal dollar in the building (or industry or transportation) sector • Elimination of U.S. consumption of Class II ozone-depleting substances, measured in tons per year of ozone-depleting potential • Reductions in melanoma and nonmelanoma skin cancers, measured by millions of skin cancer cases avoided • Percentage reduction in equivalent effective stratospheric chlorine loading rates, measured as percent change in parts per trillion of chlorine per year • Cost (industry and EPA) per ozone depletion-potential-ton phase-out targets
NASA	• As validated by external review, and quantitatively where appropriate, demonstrate the ability of NASA developed data sets, technologies, and models to enhance understanding of the Earth system, leading to improved predictive capability in each of the six science focus area roadmaps • Continue to develop and deploy advanced observing capabilities and acquire new observations to help resolve key [Earth system] science questions; progress and prioritization validated periodically by external review • Progress in understanding solar variability's impact on space climate or global change in Earth's atmosphere • Progress in developing the capability to predict solar activity and the evolution of solar disturbances as they propagate in the heliosphere and affect the Earth
NOAA	• U.S. temperature forecast skill • Determine actual long-term changes in temperature (or precipitation) throughout the contiguous United States • Reduce error in global measurement of sea surface temperature • Assess and model carbon sources and sinks globally • Reduce uncertainty in magnitude of North American carbon uptake • Reduce uncertainty in model simulations of the influence of aerosols on climate • New climate observations introduced • Improve society's ability to plan and respond to climate variability and change using NOAA climate products and information (number of peer-reviewed risk and impact assessments or evaluations published and communicated to decision makers)

continued

TABLE 2.5 Continued

Agency	Performance Measure[a]
USGS	• Percentage of nation with land-cover data to meet land-use planning and monitoring requirements (2001 nat'l data set—66 mapping units across the country) • Percentage of nation with ecoregion assessments to meet land-use planning and monitoring requirements (number of completed ecoregion assessments divided by 84 ecoregions) • Percentage of the nation's 65 principal aquifers with monitoring wells that are used to measure responses of water levels to drought and climatic variations

NOTE: DOE = Department of Energy; EPA = Environmental Protection Agency; NASA = National Aeronautics and Space Administration; USGS = U.S. Geological Survey.
[a] All are long-term measures (several years or more in the future) published with the FY 2006 budget, see <http://www.whitehouse.gov/omb/budget/fy2006/part.html>.

agers in particular continue to have difficulty establishing meaningful outcome measures, collecting timely and useful performance information, and distinguishing between results produced by the government and results caused by external factors or players such as grant recipients.[16] The report also found that issues within the purview of many agencies (e.g., the environment) are not being addressed in the GPRA context. Agency strategic plans generally contain few details on how agencies are cooperating to address common challenges and achieve common objectives. An OMB presentation to the committee acknowledged the difficulty of taking a cross-cutting view of programs such as the CCSP and identified areas in which performance measures would be especially useful.[17] These include reducing uncertainty and improving predictability; assessing trade-offs between different program elements, such as making new measurements and analyzing existing data; and demonstrating that decision support tools are helping decision makers make better choices. These issues are discussed in the following chapters.

Applicability to the CCSP

It is difficult to extrapolate performance measures from a focused agency program to the CCSP. Some agency goals overlap with CCSP goals

---

[16]General Accounting Office, 2004, *Results-Oriented Government: GPRA Has Established a Solid Foundation for Achieving Greater Results*, GAO-04-38, Washington, D.C., 269 pp.
[17]Presentation to the committee by J. Rothenberg, White House Office of Management and Budget, on March 4, 2004.

(e.g., NOAA and CCSP climate variability goals), but an agency's performance measures emphasize its mission and priorities. Moreover, annual GPRA measures are not always suitable for the long time frame required for climate change research. The PART measures allow a long-term focus, but they concentrate on limited parts of the program. The climate change PART measures (Table 2.5), for example, miss a number of CCSP priority areas (e.g., global water cycle, ecosystem function, human contributions and responses, decision support) and other important aspects of the program (e.g., strategic planning, resource allocation). Finally, agency performance measures are not designed to take account of contributions from other agencies. As a result, the aggregate of agency measures does not address the full scope of the CCSP.[18]

Nevertheless, approaches that agencies have taken to develop performance measures may be useful to the CCSP. Performance measures developed for climate change programs in the agencies provide a starting point for developing CCSP-wide metrics, and OMB guidelines and the Washington Research Evaluation Network (WREN)[19] provide tips and examples for developing metrics that are relevant to the program, promote program quality, and evaluate performance effectively (see Box A.1, Appendix A). Finally, all federal agencies with science programs rely on peer and/or internal review to evaluate research performance. Such evaluation will be especially challenging for the CCSP because of (1) a limited pool of fully qualified reviewers for multidisciplinary issues; (2) conflicts of interest, especially for experts funded by participating agencies; and (3) the high cost of conducting peer review in an era of shrinking federal budgets.

EVALUATING THE OUTCOME OF RESEARCH

Research outcomes and impacts can often be assessed only decades after the research is completed. A number of studies have attempted to trace research to outcomes, including the development of weapons systems and technological innovations, and the advancement of medicine.[20] More recently, retrospective review has become an important tool for determining whether

---

[18]Other reasons that simple performance measures cannot be aggregated across fields of research are discussed in Cozzens, S.E., 1997, The knowledge pool: Measurement challenges in evaluating fundamental research programs, *Evaluation and Program Planning*, 20, 77–89.

[19]See <http://www.science.doe.gov/sc-5/wren/>.

[20]For example, see Gibbons, M., and R. Johnston, 1974, The roles of science in technological innovation, *Research Policy*, 3, 220–242; Sherwin, C.W., and R.S. Isenson, 1967, Project Hindsight, *Science*, 156, 1571–1577; Illinois Institute of Technology Research Institute, 1968, *Technology in Retrospect and Critical Events in Science*, National Science Foundation, Washington, D.C., 2 vols.; Comroe, J.H. Jr., and R.D. Dripps, 1976, Scientific basis for the support of biomedical science, *Science*, 192, 105–111.

research investments were well directed, efficient, and productive (i.e., through the R&D investment criteria; see Appendix A), thus instilling confidence in future investments. Below is a review of the stratospheric ozone program of the 1970s and 1980s, which offers an opportunity to determine what factors made this multiagency program successful.

Lessons Learned from Stratospheric Ozone Depletion Research

The existence of ozone at high altitude and its role in absorbing incoming ultraviolet (UV) light and heating the stratosphere were deduced in the late nineteenth century. By the early 1930s, the oxygen-based chemistry of ozone production and destruction had been described.[21] However, the amount of ozone measured by instruments carried on high-altitude rockets in the 1960s and 1970s was less than expected from the reactions involving oxygen chemistry alone. Consequently, the search began for other reactive species, including free radicals, that could reduce predicted concentrations of stratospheric ozone. Among the candidate radicals considered were the NO_x group (nitric oxide and nitrous oxide), which is produced by stratospheric decomposition of nitrous oxide,[22] and the ClO_x group (chlorine atoms and ClO), which has a natural source from volcanoes and ocean phytoplankton.[23] Both radicals are also produced from rocket exhaust, which led to public concern over the possibility that space shuttle or supersonic aircraft flights in the stratosphere could lead to depletion of stratospheric ozone.

Independently, F. Sherwood Rowland and Mario Molina were investigating the fate of chlorofluorocarbon (CFC) compounds released to the atmosphere. The very unreactivity of CFCs that made them ideal refrigerants and solvents ensured that they would persist and accumulate in the atmosphere.[24] Research showed that destruction of CFCs ultimately takes place only after they are transported to high altitudes in the stratosphere, where high-energy ultraviolet photons dissociate CFC molecules and produce free chlorine radicals. On becoming aware of other work showing that

---

[21]Chapman, S., 1930, On ozone and atomic oxygen in the upper atmosphere, *Philosophical Magazine and Journal of Science*, 10, 369–383.

[22]Crutzen, P.J., 1970, The influence of nitrogen oxide on the atmospheric ozone content, *Quarterly Journal of the Royal Meteorological Society*, 96, 320–325; Johnston, H.S., 1971, Reduction of stratospheric ozone by nitrogen oxide catalysts from supersonic transport exhaust, *Science*, 173, 517.

[23]Stolarski, R.S., and R.J. Cicerone, 1974, Stratospheric chlorine: A possible sink for ozone, *Canadian Journal of Chemistry*, 52, 1610–1615.

[24]The work was later published in Lovelock, J.E., R.J. Maggi, and R.J. Wade, 1973, Halogenated hydrocarbons in and over the Atlantic, *Nature*, 241, 194–196.

chlorine atoms can catalyze the conversion of stratospheric ozone to O_2,[25] Rowland and Molina concluded that an increase in the chlorine content of the stratosphere would reduce the amount of stratospheric ozone, which in turn would increase the penetration of UV radiation to the Earth's surface.[26]

Coincident with the publication of their conclusions in *Nature* on June 28, 1974,[27] the two scientists held a press conference, although widespread press attention occurred only when they presented their results at an American Chemical Society meeting later that year. Public interest in the problem, including calls to ban the use of CFCs as propellants in aerosol spray cans, followed. The U.S. government's initial response was to create the interagency Federal Task Force on Inadvertent Modification of the Stratosphere and to commission a National Research Council (NRC) study of the problem. Reports of these groups supported the overall scientific conclusions.[28] The decision to ban CFCs in spray cans in the United States was announced in 1976 and took effect in 1978.

Subsequent scientific investigation improved understanding of the chemistry of chlorine in the stratosphere, including the formation of reservoirs such as chlorine nitrate that were not considered in earlier calculations.[29] Models used to predict future changes in stratospheric ozone, which included the HO_x, NO_x, and ClO_x chemistries, began to include more complex descriptions of the circulation of air in the stratosphere, interactions with a greater number of molecular species, and improved values (including temperature dependence) of rate constants. As a result, the magnitude of the overall effects of CFCs on stratospheric ozone predicted by the models changed. In fact, in an NRC report issued in 1984, just prior to the discovery of the Antarctic ozone hole, even the sign of ozone change was in doubt.[30]

---

[25]For example, see Stolarski, R.S., and R.J. Cicerone, 1974, Stratospheric chlorine: A possible sink for ozone, *Canadian J. Chemistry*, 52, 1610–1615; Wofsy, S.C., and M.B. McElroy, 1974, HO_x, NO_x, and ClO_x: Their role in atmospheric photochemistry, *Canadian Journal of Chemistry*, 52, 1582–1591.

[26]Molina, M.J., and F.S. Rowland, 1974, Stratospheric sink for chlorofluoromethanes: Chlorine atom-catalysed destruction of ozone, *Nature*, 249, 810–812.

[27]Molina, M.J., and F.S. Rowland, 1974, Stratospheric sink for chlorofluoromethanes: Chlorine atom-catalysed destruction of ozone, *Nature*, 249, 810–812.

[28]Federal Task Force on Inadvertent Modification of the Stratosphere, 1975, *Fluorocarbons and the Environment*, Council on Environmental Quality, U.S. Government Printing Office, Washington, D.C., 109 pp.; National Research Council, 1976, *Halocarbons: Effects on Stratospheric Ozone*, National Academy Press, Washington, D.C., 352 pp.; National Research Council, 1976, *Halocarbons: Environmental Effects of Chlorofluoromethane Release*, National Academy Press, Washington, D.C., 125 pp.

[29]National Research Council, 1984, *Causes and Effects of Changes in Stratospheric Ozone: Update 1983*, National Academy Press, Washington, D.C., 340 pp.

[30]National Research Council, 1984, *Causes and Effects of Changes in Stratospheric Ozone: Update 1983*, National Academy Press, Washington, D.C., 340 pp.

The Antarctic ozone hole, discovered serendipitously during routine monitoring of ozone levels by the British Antarctic Survey,[31] was not predicted by any model. However, work on stratospheric chemistry during the preceding decade enabled rapid deployment of tools and instruments for elucidating the cause of rapid springtime Antarctic ozone loss. Within two years the causes of ozone depletion in the Antarctic polar vortex and the impact of similar chemistry in the northern high latitudes had been determined.[32] International regulation, including the Vienna Convention (1985), the Montreal Protocol (1987), and subsequent amendments (London 1990, Copenhagen 1992, Montreal 1997, and Beijing 1999), accompanied these discoveries. Today, the response of governments to regulate stratospheric ozone depletion is viewed as a policy success, and concentrations of CFCs have leveled off or begun to decline, although it will be many decades before the Antarctic ozone hole is expected to disappear.[33]

Applicability to the CCSP

A number of lessons can be drawn from the ozone example above:

1.   **The unpredictable nature of science.** Since World War II, the U.S. government has supported a wide range of science activities because it is not possible to predict what research will turn out to be important.[34] Rowland and Molina's inquiry into the fate of a man-made chlorofluoromethane was outside the scientific mainstream, but led to a key breakthrough in the emerging field of stratospheric chemistry. (No one would have thought that the use of underarm deodorant in spray cans could influence anything at a global scale, and it is doubtful a research proposal stating so would have been funded at the time.) The Antarctic ozone hole was unpredictable in the early 1980s because the appropriate two-dimensional models with stratospheric chemistry parameters were not yet developed and key reactions (even key compounds) were not yet known. The application of these models and research had to await the independent observation of the ozone hole.

---

[31]Farman, J.C., B.G. Gardiner, and J.D. Shanklin, 1985, Large losses of total ozone in Antarctic reveal seasonal ClO_x/NO_x interactions, *Nature*, 315, 207–210.

[32]World Meteorological Organization, 1988, *Report of the International Ozone Trends Panel: 1988*, World Meteorological Organization, Report 18, Geneva, 2 vols.

[33]World Meteorological Organization, 2002, *Scientific Assessment of Ozone Depletion: 2002*, WMO Report 47, Geneva, 498 pp.

[34]Bush, V., 1945, Science, the Endless Frontier: A Report to the President, U.S. Government Printing Office, <http://www.nsf.gov/od/lpa/nsf50/vbush1945.htm>. The report led to the creation of the National Science Foundation to support research in medicine, physical and natural science, and military matters.

2. **The role of serendipity.** A dramatic loss of ozone in the lower Antarctic stratosphere was first noticed by a research group from the British Antarctic Survey that was monitoring the atmosphere using a ground-based network of instruments.[35] The same decline was famously missed by satellite observations at first because "anomalously low" values for total column ozone were flagged as potentially unreliable, and the satellite team's foremost concern at the time was its ability to accurately measure column ozone with the instrument. Subsequent reanalysis of the satellite data corroborated the existence of the Antarctic ozone hole.

3. **The role of leadership.** Aside from the initial press release by Rowland and Molina in 1974, Rowland's efforts to publicize the implications of their results were assisted by actions initiated by others, for example, the publicity department of the American Chemical Society and the politicians who called for further investigation. The resulting series of newspaper articles and interviews helped speed political outcomes, including the regulated reduction of CFCs. The rapidity of scientific progress on the causes of the Antarctic ozone hole is attributed by many involved to the leadership provided by Robert Watson, a National Aeronautics and Space Administration program manager who had both a thorough knowledge of the research he was supporting and the political awareness to release results at the most effective times.

4. **"Reduction in uncertainty."** This would have been a poor metric for evaluating scientific progress in the early stages of ozone research. Between 1975 and 1984, improved understanding and modeling of how mixtures of gases behave in the stratosphere actually increased uncertainty about the magnitude and even the sign of predicted trends in stratospheric ozone (see Figure 2.2).

5. **Role of assessments.** "State of the science" assessments can be useful for summarizing complex problems in a way that is useful to policy makers.[36] However, their usefulness in guiding future research is less clear. For example, despite the recommendations of committees convened in the 1970s and 1980s, scientific progress was not coordinated. Instead, progress was made by scientists from different fields working on the problem independently and (importantly) communicating their results broadly.

6. **Parallels with the problem of climate change are limited.** The ozone

---

[35]Farman, J.C., B.G. Gardiner, and J.D. Shanklin, 1985, Large losses of total ozone in Antarctic reveal seasonal ClO_x/NO_x interactions, *Nature*, 315, 207–210.

[36]For example, see Federal Task Force on Inadvertent Modification of the Stratosphere, 1975, *Fluorocarbons and the Environment*, Council on Environmental Quality, U.S. Government Printing Office, Washington, D.C., 109 pp.; World Meteorological Organization, 1988, *Report of the International Ozone Trends Panel: 1988*, World Meteorological Organization, Report 18, Geneva, 2 vols.

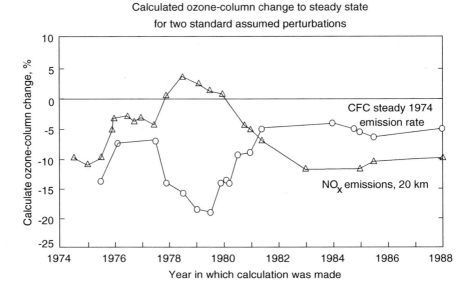

Calculated ozone-column change to steady state
for two standard assumed perturbations

FIGURE 2.2 Predictions of ozone column depletions from the same assumed chlorine and nitrogen scenarios as a function of the year for which the model was current. New discoveries in chemistry and the incorporation of better values for rate constants led to substantial fluctuations in the predictions in the late 1970s and early 1980s. SOURCE: Donald Wuebbles, University of Illinois; used with permission.

problem, although complex, involves transport and reactions in the atmosphere of a suite compounds resulting largely from human activities. Climate change, in contrast, involves a number of atmospheric trace gases, aerosols, and clouds, each of which has important cycles that are independent of human activity. Understanding the ozone hole required advances in understanding the physics of atmospheric circulation and heterogeneous chemical processes, the development of methods to measure and monitor chemical species in the stratosphere, and the modeling of feedback mechanisms. Similar progress in understanding basic physical and chemical properties of the Earth system is required before credible climate change predictions can be made. However, the scope of needed advances is vast because most greenhouse gases have important sources and sinks in the biosphere and hydrosphere, and the controls on these fluxes feed back to atmospheric composition and climate.

Finally, the Montreal Protocol and subsequent policies involve a relatively small suite of compounds. In contrast, responses to climate change could involve regulating substances important to every sector of the economy. Reductions in one greenhouse gas may be offset by increases in others. For example, increased storage of carbon in fertilized agricultural fields may be offset by increased release of nitrous oxide.

CONCLUSIONS

Although industry, academia, and federal agencies have not had to develop metrics for programs as complex as global change, their experience can provide useful guidance to the CCSP. For example, the academic experience illustrates the importance of expert judgment and peer review, which are also applicable to basic research in industry and government. The government experience (including the ozone example) shows the importance of leadership and the pitfalls of relying on a single metric such as uncertainty. Finally, the attributes of useful metrics and a methodology for creating them can be gleaned from the industry experience.

However, CCSP differs from industry in two important way that are relevant to the creation of metrics. First, in industry a manager or small management team identifies the metrics. In contrast, the CCSP program office will have to arbitrate among 13 independent agencies to choose the few important measures for guiding the program. Each of these agencies might stress a part of the program that best fulfills its mission, but would also be responsible for implementing CCSP metrics.

Second, industry operates within a framework of defined income and expenses and specific products. An increase or decrease in profits provides both a motivation to develop effective metrics and an independent check on their success. Government agencies, on the other hand, are funded by taxpayers, and frequently the "profit" is new knowledge or an innovation that is difficult to measure. Moreover, there are no simple independent checks on whether government performance measures are succeeding. A commitment by the CCSP's senior leadership to achieve and maintain outstanding performance, an open process for developing metrics, and input and feedback from outside experts and advisory groups will be required to overcome these problems.

These and other lessons are useful to guide thinking on how and why metrics should be developed and applied. Principles that can be derived from these lessons are discussed in Chapter 3.

3

Principles for Developing Metrics

*M*etrics are tools for supporting actions that allow programs to evolve toward successful outcomes, promote continuous improvement, and enable strategic decision making. Based on the lessons learned from industry, academia, and federal agencies discussed in the previous chapter, the committee offers a set of general principles to guide the development and use of metrics. Although targeted to the Climate Change Science Program (CCSP), many of these general principles have also been proposed elsewhere.[1] The principles are divided into three categories: (1) prerequisites for using metrics to promote successful outcomes, (2) characteristics of useful metrics, and (3) challenges in the application of metrics.

---

[1]For example, variations on principles 2, 3, 4, 6, 10, and 11 appear in National Science and Technology Council, 1996, Assessing Fundamental Science, <http://www.nsf.gov/sbe/srs/ostp/assess/start.htm>; principle 7 appears in Creech, B., 1994, *The Five Pillars of TQM: How to Make Total Quality Management Work for You*, Truman Talley Books, New York, 530 pp.; principles 4 and 9 are captured in Geisler, E., 1999, The metrics of technology evaluation: Where we stand and where we should go from here, Presentation at the 24th Annual Technology Transfer Society Meeting, July 15–17, 1999, <http://www.stuart.iit.edu/faculty/workingpapers/technology/>; and the importance of leadership (principle 1) appears in National Research Council, 1999, *Evaluating Federal Research Programs: Research and the Government Performance and Results Act*, National Academy Press, Washington, D.C., 80 pp.

PREREQUISITES FOR USING METRICS TO PROMOTE SUCCESSFUL OUTCOMES

1. Good leadership is required if programs are to evolve toward successful outcomes.

Good leaders have several characteristics. They are committed to progress and are capable of articulating a vision, entraining strong participants, promoting partnerships, recognizing and enabling progress, and creating institutional and programmatic flexibility. Good leaders facilitate and encourage the success of others. They are vested with authority by their peers and institutions, through title, an ability to control resources, or other recognized mechanisms. Without leadership, programmatic resources and research efforts cannot be directed and then redirected to take advantage of new scientific, technological, or political opportunities. Metrics, no matter how good, will have limited use if resources cannot be directed to promote the program vision and objectives established by the leader.

2. A good strategic plan must precede the development of metrics.

Metrics gauge progress toward achieving a stated goal. Therefore, they are meaningless outside the context of a plan of action. The strategic plan must include the intellectual framework of the program, clear and realizable goals, a sense of priorities, and coherent and practical steps for implementation. The best metrics are designed to assess whether the effort and resources match the plan, whether actions are directed toward accomplishing the objectives of the plan, and whether the focus of effort should be altered because of new discoveries or new information. Metrics, no matter how good, will have limited use if the strategic plan is weak.

CHARACTERISTICS OF USEFUL METRICS

3. Good metrics should promote strategic analysis. Demands for higher levels of accuracy and specificity, more frequent reporting, and larger numbers of measures than are needed to improve performance can result in diminishing returns and escalating costs.

Preliminary data or results are often good enough to make strategic decisions; additional effort to make them scientifically rigorous might be wasted. Larger numbers of metrics may also promote inefficiencies. For example, if a substantial amount of signed paperwork is required to demonstrate that the federal Paperwork Reduction Act is working then the metric clearly fails to meet its primary objectives.

The frequency of assessment should reflect the needs and goals of the program. Very infrequent assessments are not likely to be useful for managing programs, and overly frequent assessments have the potential to promote micromanagement or to become burdensome. For example, the Intergovernmental Panel on Climate Change (IPCC) assessments are nearly continuous and require an enormous, sustained effort by a large segment of the climate science community.[2] For short-term programs, such as the Tropical Ocean Global Atmosphere (TOGA) experiment, frequent scientific assessments would have been nearly useless, because a decade was required to clearly demonstrate some of the most important scientific outcomes.[3] On the other hand, process metrics for evaluating progress on the creation and operation of the program, would have had value on much shorter time scales.

4.   Metrics should serve to advance scientific progress or inquiry, not the reverse.

A good metric will encourage actions that continuously improve the program, such as the introduction of new measurement techniques, cutting-edge research, or new applications or tools. On the other hand, a poor measure could encourage actions to achieve high scores (i.e., "teaching to the test") and ultimately unbalance the research and development portfolio. The misapplication of metrics could lead to unintended consequences, as illustrated by the following examples:

• The author citation index provides a measure of research productivity. If this metric were the only way to measure faculty performance, it could drive researchers to invest more in writing review articles that are cited frequently than in working on new discoveries.

---

[2]The IPCC was established in 1988 under the auspices of the United Nations Environment Programme and the World Meteorological Organization to conduct assessments of climate change and its consequences. Assessments are produced and peer reviewed by more than 1000 scientific researchers, policy experts, and risk analysts from all over the world. Writing and review of the assessment reports, which have been produced about every five years since 1990, take several years.

[3]The development of the TOGA program, establishment of the TOGA-TAO (Tropical Atmosphere Ocean) observation array, and demonstration that the improved observations and process studies promoted improved forecasting and understanding of El Niño-Southern Oscillation (ENSO) events required more than a decade of effort. See National Research Council, 1996, *Learning to Predict Climate Variations Associated with El Niño and the Southern Oscillation: Accomplishments and Legacies of the TOGA Program*, National Academy Press, Washington, D.C., 171 pp.

• The U.S. Global Change Research Program (USGCRP) has supported efforts to compare major climate models. Convergence of model results (e.g., similar temperature increases in response to a doubling of carbon dioxide) could be a measure of progress in climate modeling. The metric succeeds if it identifies differences in the way physical processes are incorporated in models, which then leads to research aimed at improving understanding of those processes and, eventually, to model improvements and the reduction of uncertainties in model predictions. The metric fails if it creates an unintended bias in researchers who adjust their models solely to bring them into better agreement with one another.

5. Metrics should be easily understood and broadly accepted by stakeholders. Acceptance is obtained more easily when metrics are derivable from existing sources or mechanisms for gathering information.

It is important to avoid creating requirements for measurements that are difficult to obtain or that will not be recognized as useful by stakeholders. The latter is especially difficult for innovative or multidisciplinary sciences that have yet to establish natural mechanisms of assessment. The following examples illustrate these points:

• A metric for measuring change in forest cover is the fraction of land surface covered by forest canopy, which is detectable using remote sensing. An area is considered "forest" when 10 to 70 percent of the land surface is covered by canopy. However, the lower threshold would not be viewed as useful by stakeholders. A metric based on this threshold (essentially, forest or not forest) could mean that an area with dense canopy would be defined as forest, despite being severely logged and degraded.[4] The metric becomes more useful when it is associated with information about land-cover types. For example, a 10 percent threshold might be appropriate for savannah areas, whereas higher thresholds would be required for ecosystems with more continuous canopy cover. More detailed measures of forest cover can also be developed, such as selective removal of specific tree types, changes in species composition, or changes in indices (e.g., seed production, primary productivity, leaf density). However, one can quickly reach a point at which the difficulty of measuring the quantities systematically becomes overwhelming, limiting their use as metrics.

---

[4]Intergovernmental Panel on Climate Change, 2000, *Land Use, Land-Use Change, and Forestry: A Special Report*, R.T. Watson, I.R. Noble, B. Bolin, N.H. Ravindranath, D.J. Verardo, and D.J. Dokken, eds., Cambridge University Press, Cambridge, U.K., p. 375.

• The number of users is commonly cited as a metric of the useful-ness of holdings in data centers.[5] However, it is difficult to gather reliable information to support this metric. With the shift to on-line access, most users find and retrieve data via the Internet. Since the actual number of users is not known, data centers count "hits" on their web sites, which are likely to be several orders of magnitude greater than the actual number of users, or "distinct hosts," which overcount users accessing the site from several different computers.

6. Promoting quality should be a key objective for any set of metrics. Quality is best assessed by independent, transparent peer review.

The success of the scientific enterprise and confidence in its results depend strongly on the quality of the research. Although peer review has well-known limitations (e.g., results depend on the identity of the reviewers, there is a tendency to view research results conservatively), it is the gener-ally accepted mechanism to assess research quality. Review occurs through-out the scientific enterprise in the form of peer review of proposals submitted for funding, peer review of manuscripts submitted for publication in journals, and internal and peer review of programs and program outcomes (Boxes 2.1 and 2.2). Peer review also provides the best mechanism for judging when to change research directions and, thus, make programs more evolutionary and responsive to new ideas.

7. Metrics should assess process as well as progress.

The success of any program depends on many factors, including pro-cess (e.g., level of planning, type of leadership, availability of resources, accessibility of information) and progress (e.g., addition of new observa-tions, scientific discovery and innovation, transition of research to practical applications, demonstration of societal benefit). The assessment of process as well as progress is important for every program, but its value is particu-larly high for large, complex programs.

The sheer diversity and complexity of programs such as the USGCRP and the CCSP defies the application of a few simple metrics. Even the assessment of progress depends on the nature and maturity of the effort. Enhancing an existing data set is different from developing a new way to

---

[5]National Research Council, 2003, *Review of NOAA's National Geophysical Data Center*, The National Academies Press, Washington, D.C., 106 pp.; National Research Council, 2002, *Assessment of the Usefulness and Availability of NASA's Earth and Space Science Mission Data*, National Academy Press, Washington, D.C., 100 pp.

measure a specific variable. Process studies are different from model improvements. Mission-oriented science is different from discovery science. Metrics should reflect the diversity and complexity of the program and the level of maturity of the research. Comprehensive assessment of the program will include processes taken to achieve CCSP goals, as well as progress on all aspects of the research, from inputs to outputs, outcomes, and impacts.

8.   A focus on a single measure of progress is often misguided.

The tendency to try to demonstrate progress with a single metric can create an overly simplistic and even erroneous sense of progress. Reliance on a single metric can also result in poor management decisions. These points are illustrated in the following examples:

•   The predicted increase in globally averaged temperature with a doubling of carbon dioxide has remained in the same range for more than 20 years (see Chapter 4). According to the metric of reducing uncertainty, climate models would seem to have advanced little over that period despite considerable investment of resources. In fact, however, the physics incorporated in climate models has changed dramatically. Incorporation of new processes, such as vegetation changes as a function of climate, is yielding previously unrecognized feedbacks that either amplify or dampen the response of the model to increased carbon dioxide. The result is often greater uncertainty in the range of predicted temperatures until the underlying processes are better understood. New discoveries can also indicate that certain elements of the weather and climate system are not as predictable as once thought. In such cases, significant scientific advance can result in an increase in uncertainty. Rather than relying solely on uncertainty reduction, it may be more appropriate to develop metrics for the three components of uncertainty: (1) success in identifying uncertainties, (2) success in understanding the nature of uncertainties, and (3) success in reducing uncertainties.
•   Change in biomass is commonly used as a metric to assess the health of marine fisheries. However, this metric fails to recognize the substitution of one species for another (an important indication of environmental change or degradation), interactions among species, and changes in other parts of the food web that result from fishing. Reliance on biomass alone could lead to the establishment of fishing targets that speed the decline of desirable fish stocks or adversely affect other desired species. For example, early management of Antarctic krill stocks strictly on a biomass basis did not account for two facts: (1) most harvesting was in regions that support feeding by large populations of krill-dependent predators such as penguins, whales, and seals, and (2) predator populations can be adversely affected by

krill fishing, especially during their breeding seasons.[6] A more complex metric or set of metrics that incorporate species composition (multispecies management), information about dependent species (ecosystem-based management), and species distribution and environmental structure (area-based management) would reflect the state of knowledge and lead to better resource management decisions. Combining a biomass-based metric with information from quota-based or fishing-effort-based management practices would provide an approach for sustaining fishery stocks at levels that are both economically and environmentally desired.

CHALLENGES IN THE APPLICATION OF METRICS

9. Considerable challenge should be expected in providing useful a priori outcome or impact metrics for discovery science.

The assignment of outcome metrics implies that we can anticipate specific results. This works well at the level of mission-oriented tasks such as increasing the accuracy of a thermometer. However, much of discovery science involves the unexpected and the outcome is simply unknown. For example, the measurement of atmospheric carbon dioxide concentrations by C.D. Keeling eventually revealed both an annual cycle and a decadal trend in atmospheric composition, neither of which was the original goal of the observation program.[7] This remarkable achievement could have been defeated by the strenuous application of outcome metrics aimed at determining whether a reliable "baseline" CO_2 level in the atmosphere had been established.

It is difficult to conceive of metrics for serendipity, yet serendipity has resulted in numerous discoveries—from X-rays to Post-it adhesives. Great care must be taken to avoid applying measures that stifle discovery and innovation. The most suitable metrics may be related to process (e.g., the level of investment in discovery, the extent to which serendipity is encouraged, the extent to which curiosity-driven research is supported). The National Science Foundation is highly regarded for its ability to promote discovery science, and its research performance measures focus on processes for developing a scientifically capable work force and tools to enable discovery, learning, and innovation (Table 2.4).

---

[6]For example, see Committee for Conservation of Antarctic Marine Living Resources, 2003, *Report of the 22nd Meeting of the Scientific Committee*, SC-CCAMLR-XXII, Hobart, Australia, 577 pp.

[7]Weart, S.R., 2003, *The Discovery of Global Warming*, Harvard University Press, Boston, 240 pp.

10. Metrics must evolve to keep pace with scientific progress and program objectives.

The development of metrics is a learning process. No one gets it right the first time, but practice and adjustments based on previous trials will eventually yield useful measures and show what information must be collected to evaluate them. Metrics must also evolve to keep pace with changes in program goals and objectives. Scientific enterprises experience considerable evolution as they move through various phases of exploration and understanding. Metrics for newly created science programs, which focus on data collection, analysis, and model development to increase understanding, will tend to focus on process and inputs. As the science matures and the resulting knowledge is applied to serve society, metrics will focus more on outputs and, finally, on outcomes and impacts. As science transitions from the discovery phase to the operational or mission-oriented phase, the types of metrics should also be expected to evolve.

11. The development and application of meaningful metrics will require significant human, financial, and computational resources.

The development and application of metrics, especially those that focus on quality, is far from a bookkeeping exercise. Efforts to assess programmatic plans, scientific progress, and outcomes require substantial resources, including the use of experts to carry out the reviews. Funding to support the logistics of the reviews is also required. The CCSP strategic plan includes a substantial number of assessments and a growing emphasis on measurable outcomes. As these are implemented, the choice of meaningful measures of progress must be deliberate. If the IPCC process is a representative example, the growing emphasis on assessments has the potential to increasingly divert resources from research and discovery to assessment.

4

Characterizing and Reducing Uncertainty

*T*he term *reducing uncertainty* is ubiquitous within the Climate Chance Science Program (CCSP) strategic plan. Reducing uncertainties is the central theme of one of the five major CCSP goals and the foundation of one of the four core approaches to address these goals. As such, it is viewed as a litmus test for determining whether scientific knowledge is sufficient to justify particular policies and decisions. It is listed as one of the key criteria for prioritization of work elements within the CCSP. Finally, "reducing uncertainties" appears in many of the research questions, milestones, and products and is an element of plans to develop decision support resources (Chapter 11 of the plan calls for "scientific synthesis and analytic frameworks to support integrated evaluations, including explicit evaluation and characterization of uncertainties"). The fact that the concept of reducing uncertainties appears in the plan as a goal, within an approach, as a criteria, as the basis of scientific questions, and as a milestone or product is indicative of the degree to which this concept pervades strategic thinking in the CCSP. A key question is whether reduction of uncertainty is also a metric for assessing progress, and if so, how should it be applied?

In a number of presentations to the committee, reducing uncertainty appeared to take on the mantle of a potential "supermetric" capable of assessing whether or not the CCSP is successful. For example, if the investment in climate model prediction does not result in a narrowed range of predicted sensitivity, then the investment could be viewed as a failure. This use of reducing uncertainty as a metric violates the general principles presented in Chapter 3. Reliance on a single metric can provide an erroneous sense of progress and increase the potential for misuse (principle 8). The principle that metrics should address both process and progress (principle 7) is particularly relevant for complex and diverse programs such as the CCSP. Importantly, the meaning of uncertainty is poorly defined for much of the scope of the CCSP. It is likely that different definitions apply to different program elements (e.g., overarching goal, prioritization criteria, research question, milestone). Without careful definition, reducing uncertainty cannot be evaluated using specific observable or articulated measures. Therefore, it violates the principle that metrics should be easily understood and broadly accepted by the community (principle 5). To be meaningful, a metric must first be based on a well-specified variable that indicates advancement of knowledge. Second, a precise definition of what is meant by "uncertainty" in reference to that variable must be specified. The pervasive and diverse use of reducing uncertainty as a definition of progress, and the flaws and potential misuse of reducing uncertainty as a metric, warrant a more detailed assessment of its application for the CCSP.

THE ROLE OF UNCERTAINTY IN CLIMATE DISCUSSIONS

The climate community expresses uncertainty in different ways.[1] The CCSP defines uncertainty as:

> An expression of the degree to which a value (e.g., the future state of the climate system) is unknown. Uncertainty can result from lack of information or from disagreement about what is known or even knowable. It may have many types of sources, from quantifiable errors in the data to ambiguously defined concepts or terminology, or uncertain projections of human behavior.[2]

Uncertainty plays a key role in policy formation because decisions often turn on the question of whether scientific understanding is sufficient to justify particular types of response. The CCSP strategic plan seeks to develop knowledge of the complex human-natural system in support of public and private decisions, and a central component of this task concerns character-

---

[1]For example, see Lempert, R., N. Nakicenovic, D. Sarewitz, and M. Schlesinger, 2004, Characterizing climate-change uncertainties for decision-makers, *Climatic Change*, 65, 1–9; Intergovernmental Panel on Climate Change, 2004, *Describing Scientific Uncertainties in Climate Change to Support Analysis of Risk and of Options: Workshop Report*, M. Manning, M. Petit, D. Easterling, J. Murphy, A. Partwardhan, H.-H. Rogner, R. Swart, and G. Yohe, eds., Report of a workshop held at the National University of Ireland, Maynooth, May 11–13, 2004, 138 pp., <http://ipcc-wg1.ucar.edu/meeting/URW/product/URW_Report_v2.pdf>.
[2]Climate Change Science Program and Subcommittee on Global Change Research, 2003, *Strategic Plan for the U.S. Climate Change Science Program*, Washington, D.C., p. 199.

izing, and where possible reducing, current levels of uncertainty in knowledge of key climate processes.

The CCSP objective is laudable, but it has also yielded the potential for an overly simplistic view of uncertainty as a measure of progress. Perhaps the most prominent example of this shortcoming involves comparison of the estimates of the change in global mean temperature at equilibrium with a doubling of CO_2 from preindustrial levels. Studies of global mean temperature date back as far as the late nineteenth century,[3] when Arrhenius estimated that a doubling of CO_2 would warm the planet by 4-6°C. However, the most prominent early modern estimate was provided by a 1979 National Research Council study, usually referred to as the Charney report in reference to the committee's chairman.[4] It concluded that " . . . the equilibrium surface global warming due to doubled CO_2 will be in the range 1.5°C to 4.5°C, with the most probable value near 3°C." Subsequent estimates of this range have led to similar conclusions. For example, the 1995 Intergovernmental Panel on Climate Change (IPCC) second assessment reports the same range with a "best estimate" of 2.5°C, while the 2001 third assessment states that "the previously estimated range for this quantity, widely cited as +1.5°C to +4.5°C, still encompasses the more recent model sensitivity estimates."[5] Such comparisons of ranges are widely interpreted, even in the scientific literature, as meaningful indicators of temperature change (or lack of it).[6] The application of an uncertainty metric, defined in this case as the extent to which the range in estimated climate sensitivity due to the doubling of carbon dioxide has narrowed with climate research, would suggest that little progress has been made. In fact, this interpretation is far from correct because of the flaws in the application of uncertainty as a metric.

PITFALLS IN THE APPLICATION OF UNCERTAINTY METRICS

Previous experiences in the climate debate, associated studies, and the example above reveal three circumstances in which caution should be exer-

---

[3]Arrhenius, S., 1896, On the influence of carbonic acid in the air upon the temperature of the ground, *Philosophical Magazine and Journal of Science*, 41, 251.

[4]National Research Council, 1979, *Carbon Dioxide and Climate: A Scientific Assessment*, National Academy Press, Washington D.C., 22 pp.

[5]Intergovernmental Panel on Climate Change, Working Group I, 1995, *Climate Change 1995: The Science of Climate Change*, Cambridge University Press, Cambridge, U.K., p. 34; Intergovernmental Panel on Climate Change, Working Group I, 2001, *Climate Change 2001: The Scientific Basis*, Cambridge University Press, Cambridge, U.K., p. 527.

[6]For example, see "Rising global temperature, rising uncertainty," *Science*, 292, April 13, 2001, pp. 192–194; "Three degrees of consensus," *Science*, 305, August 13, 2004, pp. 932–934.

cised in the construction of metrics for the CCSP: (1) ill-specified comparisons, (2) systematic errors, and (3) chaotic systems.

Ill-Specified Comparisons

Even in cases where a variable is well defined, improper comparisons can arise. A clear example of such failure is the comparison of the estimates in the Charney report with the most recent IPCC analysis described above. The two estimates of mean global temperature sensitivity to increased carbon dioxide cannot be compared meaningfully because neither states what confidence interval is intended. The Charney report did not include a statement about confidence intervals, and the IPCC has not attached probabilities or confidence intervals to its ranges of estimates. The same issue arises with comparisons of the estimated range of temperature change among the IPCC summaries, which involve both climate and emissions models.[7] Therefore, a comparison of the estimates from the Charney report and the more recent IPCC reports is not meaningful because what is meant by uncertainty must be determined more precisely for each case. Metrics for the CCSP will have to take these challenges explicitly into account with careful definitions of system variables and their associated uncertainties.

Correction of Systematic Errors

There is no foolproof methodology for determining systematic error. An empirically observed correlation between a presumed cause and effect may be wholly spurious due to omission of a causal factor. Finding all systematic errors necessitates examining a whole series of ad hoc possibilities, some of which may be completely unknown to the observer.

The nature of the problem is shown in Figure 4.1, which illustrates one of the most well-established areas of scientific research: the speed of light. Figure 4.1 displays the results of the well-specified problem of determining the speed of light under carefully controlled laboratory conditions at different times. In assuming that the most recent value is indeed closest to the true value (an assumption that would require serious effort to confirm), it might have been anticipated that the standard errors of the earlier measurements—made by extremely careful and expert physicists—would have encompassed the apparent correct value. That many of the earlier results do not do so likely can be attributed either to random effects or to unsuspected, and therefore uncorrected, systematic errors.

---

[7]Reilly, J., P.H. Stone, C.E. Forest, M.D. Webster, H.D. Jacoby, and R.G. Prinn, 2001, Uncertainty and climate change assessments, *Science*, 293, 430–433.

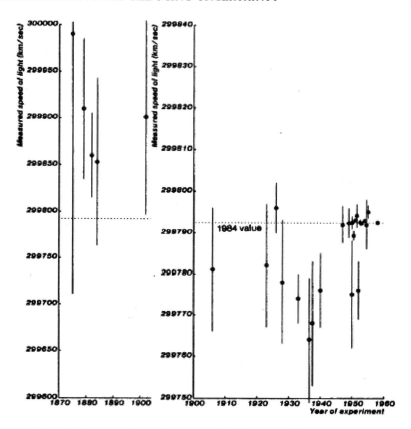

FIGURE 4.1 Estimated values of the speed of light at different points in history. Vertical bars are the expected value with standard error. Note that the vertical scales are slightly different. SOURCE: Henrion, M., and B. Fischhoff, 1986, Assessing uncertainty in physical constants, *American Journal of Physics*, **54**, 791–798. Copyright 1986, American Association of Physics Teachers.

Errors must be considered, even for a problem as simple as determining average temperature. To measure temperature accurately, it is necessary to consider the potentially erroneous calibration of the thermometer, dependence on the housing of the thermometer, urban development in the vicinity of the measurements, and differences in the way different observers read a thermometer. Realistic representation of uncertainties requires estimates of the possible contributions of all significant effects, some of which may be relatively unknown.

As science advances, phenomena often become more fully characterized. One typical result is that phenomena not initially understood to be relevant are found to contribute to the effect being forecast—perhaps increasing uncertainty as a result. One such example is the introduction of an interactive land surface into climate models in the interval of time between the publication of the Charney report and the most recent IPCC assessment. These interactive model components were added to global climate models because climate-vegetation feedbacks were discovered to be a potential mechanism for altering climate sensitivity predictions. Such an innovation does not necessarily reduce uncertainty and in fact, given the diversity of interactive land surface models, is likely to have increased uncertainty. Nevertheless, it is not a research failure, because it contributes to an advanced characterization of the problem. However, it might be erroneously classified as such by an incautious application of uncertainty reduction as the sole metric for measuring advancements in knowledge.

Revelation of Chaotic Systems

For some variables and for some scales, the Earth's climate and weather systems are chaotic. That is, in any projection of this nonlinear system, irreducible small errors in the initial conditions increase with time until the prediction becomes meaningless. Prediction beyond a certain time horizon is impossible in principle. In classic work by E. Lorenz, this phenomenon was found to hold for weather systems.[8] Consider the problem of predicting precipitation in Washington, D.C., on July 1 of any particular year. Three- and five-day forecasts of precipitation can be made with considerable skill. Fifty years ago, a meteorologist might have rationally concluded that obtaining accurate forecasts several weeks in advance would be only a matter of time and effort and that the uncertainty existing at that time could only diminish as models and observations improved. Today we know, through theories of chaos, that such an expectation would have been false— there is virtually no hope of accurate forecasts of daily precipitation one month ahead. Indeed, as observational records have grown in length and as ever-more-extreme events are recorded, estimates of the uncertainty of month-ahead precipitation forecasts have increased, not decreased. Hence, a forecast made in 1950 might well have been considerably more optimistic than one made today, but less accurate scientifically.

As knowledge of the interacting systems increases, estimates of uncertainty associated with some climate variables and some scales could decrease and/or increase, perhaps markedly.

---

[8]Lorenz, E., 1963, Deterministic nonperiodic flow, *Journal of Atmospheric Science*, 20, 130–141.

USE OF UNCERTAINTY METRICS

The ability to properly characterize uncertainty is of great value. Uncertainty plays a key role in policy formation because decisions often turn on the question of whether scientific understanding is sufficient to justify particular types of response. For this reason, metrics that mark advancements in this area will be valuable. However, it must be reemphasized that advances in the knowledge of climate systems may result not only from decreases in uncertainty, but also from increases as more is understood about governing elements. Hence, imprudent application of any simple measure of uncertainty could be very damaging to scientific efforts. In many cases, it may be more useful to consider successes in identifying uncertainties and successes in understanding the nature of uncertainties.

The problem of living with contingencies whose uncertainty cannot be reduced or eliminated is a familiar one and has led over the centuries to a practice of risk-reducing investment, insurance, and expenditure to maintain options for future choice. Global change falls into this category—possible outcomes (no global warming, moderate to severe global warming, global cooling) can be stated only in probabilistic terms.[9]

Given the constraints described above, reduction of uncertainty should not be relied upon as a metric for assessing progress in the CCSP. Alternative measures that do not have these shortcomings are presented in Chapter 6.

---

[9]Mastrandrea, M.D., and S. Schneider, 2004, Probabilistic integrated assessment of "dangerous" climate change, *Science*, **304**, 571–575.

5

Process of Developing Metrics

*T*he committee was asked to identify performance measures and metrics for documenting progress, measuring future performance, and communicating levels of performance in three to five areas of climate and global change research. This task presented a considerable challenge because of the enormous breadth of the Climate Change Science Program (CCSP) and the lack of models for developing metrics for multi-agency programs. This chapter describes the process by which the committee developed metrics for the CCSP.

FRAMEWORK FOR MEASURING PROGRESS TOWARD CCSP GOALS

Metrics are intended to assess progress toward stated goals. The Office of Management and Budget (OMB) notes that many strategic goals are difficult to measure and encourages agencies to develop "specific, operational performance goals that align with strategic goals."[1] Performance measures are then created for these operational performance goals.

---

[1]Office of Management and Budget, 2005, Guidance for Completing the Program Assessment Rating Tool (PART), p. 9, <http://www.whitehouse.gov/omb/part/fy2005/2005_guidance.doc>.

OMB Approach	CCSP	Committee Approach
Strategic goals	Strategic goals	CCSP strategic goals
Performance goals	Questions	
	Milestones, products, payoffs	Eight themes
Performance measures		Metrics

The CCSP strategic plan does not contain operational performance goals, and the committee found that the five CCSP goals are indeed stated in terms that are too broad to serve as a framework for developing meaningful metrics. Consequently, the committee considered the 224 milestones, products, and payoffs identified in the CCSP strategic plan, which provide greater specificity about what the program is trying to achieve. The committee found that the milestones, products, and payoffs could be grouped into eight themes for which metrics could be developed. These themes are

1. improve data sets in space and time (e.g., create maps, databases, and data products; densify data networks);
2. improve estimates of physical quantities (e.g., through improvement of a measurement);
3. improve understanding of processes;
4. improve representation of processes (e.g., through modeling);
5. improve assessment of uncertainty, predictability, or predictive capabilities;
6. improve synthesis and assessment to inform;
7. improve the assessment and management of risk; and
8. improve decision support for adaptive management and policy making.

The phrasing of these eight themes either matches or is closely allied with the phrasing of nearly all of the program's milestones, products, and payoffs. In addition, the themes represent a sequence in scientific investigation, starting from the development of new or better observations, to an improved understanding of processes, to an improved capability to predict or forecast future climate changes, and finally to improved use of information to better serve society. As such, they offer an organizing framework for developing metrics for assessing the full range of CCSP activities.

The committee proceeded under the assumption that metrics would be very different for each of these themes and that developing quantifiable measures for many elements of the CCSP would be difficult. For example, metrics to assess improvements in CO_2 observing systems seemed likely to differ from metrics to evaluate new knowledge about processes that control carbon sources and sinks. Moreover, it seemed likely that metrics might be specific to a particular program element, such as understanding climate change feedbacks.[2]

To test these assumptions, the committee chose one or two case studies for each of the eight themes, based on the areas of expertise of the members. Each case study followed the same format: (1) an introduction to the issue being assessed; (2) a description of the relevant milestone, product, or payoff being addressed as stated in the CCSP strategic plan; and (3) example metrics divided into the categories of process, input, output, outcome, and impact metrics. In addition, the difficulty of assigning the metrics to the case study objectives and generalizing them to other parts of the CCSP was assessed. Two case studies illustrating different kinds of CCSP objectives are presented in the next section, and a selection of others, in the preliminary form that guided the committee, is summarized in Appendix B.

The case studies were also mapped onto the CCSP overarching goals and themes as a second check on the breadth of analysis provided by this approach (Table 5.1).

EXAMPLE CASE STUDIES

Two examples illustrate the differences and similarities of metrics developed for very different parts of the program. The first—the effect of carbon dioxide on land carbon balance—is science oriented and illustrates the CCSP theme of improving understanding of processes (theme 3). The second—adaptive management of water resources—is oriented toward decision support (theme 8). In developing the respective metrics, Tables 5.2 and 5.3, the committee considered relevant performance measures from agency strategic plans and Program Assessment Rating Tool (PART) submissions (e.g., Tables 2.4 and 2.5), generic research and development (R&D) metrics developed elsewhere (Appendix C), the literature, and the committee's experience with the CCSP research question.

---

[2]Metrics for the latter were proposed in National Research Council, 2003, *Understanding Climate Change Feedbacks*, The National Academies Press, Washington, D.C., 152 pp. They include (1) comparison of observed and simulated response of clouds, water vapor, and lapse rate to every well-observed forcing mechanism and time scale, including the diurnal and seasonal response, the response to ENSO (El Niño-Southern Oscillation), and the response to volcanic eruptions; (2) the accuracy with which Earth system models can reproduce observed diurnal and seasonal variations of the hydrological cycle over land; (3) total water column heat content along decadally monitored transoceanic cross sections; and (4) tropical Pacific sea surface temperature and pycnocline depth to evaluate model performance and to diagnose and monitor decadal and longer-term changes in ENSO statistics.

TABLE 5.1 Relationship Between the CCSP Question That the Case Study (a-j) Is Trying to Address, CCSP Overarching Goals, and Committee-Identified Themes

Themes	CCSP Overarching Goals				
	Improve Knowledge	Improve Quantification	Reduce Uncertainty	Understand Adaptability	Manage Risk
Improve data sets	*a, b*	*a, b*	*b*		
Estimate physical quantities	*c*	*c*	*c*		
Understand processes	*d*	*d*			
Represent processes	*e*	*e*	*e*	*e*	
Assess uncertainty, predictability	*f*	*f*	*f*	*f*	
Synthesize and assess to inform	*g*	*g*	*g*	*g*	*g*
Assess and manage risk	*h*	*h*	*h*	*h*	*h*
Adaptive management, policy making	*i*		*i, j*	*i, j*	*i, j*

KEY:

a = Case study on solar forcing of climate: To what extent are climate changes as observed in instrumental and paleoclimate records related to volcanic and solar variability, and what mechanisms are involved in producing climate responses to these natural forcings?

b = Case study on aerosols and their role in climate forcing: What are the climate-relevant chemical, microphysical, and optical properties, and the spatial and temporal distributions, of human-caused and naturally occurring aerosols?

c = Case study on sea-level rise: What are the projected contributions from different components of the climate system to future sea-level changes, what are the uncertainties in the projections, and how can they be reduced?

d = Case study on the effect of carbon dioxide on land carbon balance: What are the potential consequences of global change for ecological systems?

e = Case study on climate-vegetation feedbacks: What are the most important feedbacks between ecological systems and global change (especially climate), and what are their quantitative relationships?

TABLE 5.1 Continued

f = Case study on paleoclimate time series as benchmarks of climate variability and change: To what extent are climate changes as observed in instrumental and paleoclimate records related to volcanic and solar variability, and what mechanisms are involved in producing climate responses to these natural forcings?

g = Case study on human health and climate: What are the potential human health effects of global environmental change, and what climate, socioeconomic, and environmental information is needed to assess the cumulative risk to health from these effects?

h = Case study on assessing, preventing, and managing public health threats of infectious diseases: How can the methods and capabilities for societal decision making under conditions of complexity and uncertainty about global environmental variability and change be enhanced? What are the potential human health effects of global environmental change, and what climate, socioeconomic, and environmental information is needed to assess the cumulative risk to health from these effects?

i = Case study on adaptive management of water resources: How can information on climate variability and change be most efficiently developed, integrated with nonclimatic knowledge, and communicated in order to best serve societal needs?

j = Case study on policy making based on scenarios of greenhouse emissions and climate response: What are the current and potential future impacts of global environmental variability and change on human welfare, what factors influence the capacity of human societies to respond to change, and how can resilience be increased and vulnerability decreased?

Effect of Carbon Dioxide on Land Carbon Balance

Related CCSP Questions, Milestones, and Products. Question 8.2: "What are the potential consequences of global change for ecological systems?"[3] The related milestones, products, and payoffs include improved understanding of processes about (1) how elevated CO_2 concentrations, warming, and altered hydrology will influence the productivity of land plants and the net carbon balance of terrestrial ecosystem; and (2) how this response will evolve over time in response to other factors that influence carbon storage in ecosystems over the next century.

Rationale. The capacity for CO_2 "fertilization" in land ecosystems may be responsible for some of the apparent land carbon sink observed in the 1990s.[4] However, the means by which the products of increased photosynthesis are allocated and the fate (and therefore residence time) of plant

---

[3]Climate Change Science Program and Subcommittee on Global Change Research, 2003, *Strategic Plan for the U.S. Climate Change Science Program*, Washington, D.C., pp. 87–89.

[4]Schimel, D.S., Terrestrial ecosystems and the carbon cycle, *Global Change Biology*, 1, 77–92, 1995; Intergovernmental Panel on Climate Change, Working Group I, 2001, *Climate Change 2001: The Scientific Basis*, Cambridge University Press, Cambridge, U.K., pp. 195–196; Schimel, D.S., and 29 coauthors, 2001, Recent patterns and mechanisms of carbon exchange by terrestrial ecosystems, *Nature*, 414, 169–172.

TABLE 5.2 Effect of CO_2 on Land Carbon Balance

Type	Example Metrics
Process	• Has the leadership of this overall effort, which spans several agencies, been identified? • Does a structure exist that will involve the scientific community in planning the sites and conditions chosen for manipulation or gradient studies? • Is there a 5-10-year plan for implementation of the manipulation experiments, to be revisited and updated in accord with new discoveries? • Is there a plan to incorporate longer-term aspects of the problem that extend beyond the 5-10-year horizon (i.e., multiple generations of plants exposed to altered atmospheric conditions)? • Do a mechanism and timetable exist for periodic review of experimental implementations, including testing of model predictions outside experimental areas? • Do a mechanism and timetable exist to disseminate results to potential stakeholders (particularly the agricultural community) and involve them in planning discussions?
Input	• Is there sufficient theoretical basis for the design and interpretation of experiments? • Is the technology available to perform experiments assuming multiple, long-term (decadal) manipulations of plots of sufficient size to test hypotheses? • Are sufficient resources (people, dollars) available to implement and support a measurement network, modeling, and interpretive activities for the appropriate period of time (decades)? • Is there an identified stakeholder community to take advantage of scientific advances?
Output	• Peer-reviewed, published results generated for each site and synthesis activities across sites that identify the most important mechanisms at work • Production of a facility that (1) can be put into the field for years at a time and (2) can maintain atmospheric CO_2 levels at a specific set point (e.g., 50 ppm [parts per million] above ambient levels), with a precision (averaged over 1 hour) of 5 ppm. For a subset of these systems, additional control over either atmospheric ozone levels, temperature (i.e., increase by 5°C compared to the control plot), soil moisture, or species diversity is required • Development of a suite of new measurement techniques that can detect carbon allocation patterns on time scales of (1) hours, (2) days to weeks, and (3) a growing season in response to external variables and photosynthetic rates of plants in control versus experimentally manipulated systems • Incorporation of relationships between photosynthetic rates, carbon allocation, and external and internal variables into process-based models that simulate patterns of photosynthetic response and allocation (on appropriate time scales for each process) and that can be tested against other observations as well as in other kinds of manipulated systems • Technology developed for rapid control of trace gas concentrations at high precision

TABLE 5.2 Continued

Type	Example Metrics
Outcome	• Peer-reviewed and published knowledge of the processes by which increasing atmospheric CO_2 can influence the carbon balance at (1) the whole plant level and (2) the ecosystem level. Determination of the sign and magnitude (to 30%) of the feedback between CO_2 levels and the amount of carbon stored over the first year of the manipulation (and subsequent years as they become available) • Models of suitable spatial scale that incorporate process-level understanding are used to predict the response of ecosystems to multiple stressors, such as increased CO_2 and temperature or CO_2 and ozone • Policy makers are informed about — The potential for different kinds of ecosystems to store or release carbon under conditions of a 50 ppm increase in atmospheric CO_2 — The magnitude of release or uptake of CO_2 and how this understanding will be modified by the presence of more investigators in the field • Peer-reviewed assessments that quantify the potential effects of changing atmospheric composition on the yield of different crops • Improved prediction of future trends in atmospheric CO_2 levels, given a scenario of fossil fuel emissions and deforestation
Impact	• Crop productivity is improved because of use of forecasts that take into account changes in CO_2, ozone, and climate • Conservation reserves are more resilient because of use of knowledge of how changes in CO_2 affect plant competition and ecosystem structure

carbon stores are still matters of debate and uncertainty.[5] On longer time scales, when factors such as disturbance frequency must be included in assessments of land carbon balance, even the sign of land carbon response to elevated CO_2 is uncertain.[6] Higher-CO_2 conditions may favor one kind of plant over another—changing the structure of ecological communities, their functions (including carbon storage), and their vulnerability to disturbance such as fire. Further complications arise when increased CO_2 is correlated with other factors that affect plant productivity, such as changes in climate, deposition of excess nitrogen, presence of high O_3 levels, or

[5]Bazazz, F.A., 1990, The response of natural ecosystems to rising global CO_2 levels, *Annual Review of Ecology and Systematics*, **21**, 167–196; Woodwell, G.M., F.T. Mackenzie, R.A. Houghton, M. Apps, E. Gorham, and E. Davidson, 1998, Biotic feedbacks in the warming of the Earth, *Climatic Change*, **40**, 495–518.

[6]Korner C., 2004, Through enhanced tree dynamics carbon dioxide enrichment may cause tropical forests to lose carbon, *Philosophical Transactions of the Royal Society of London, Series B*, **359**, 493–498; Chambers, J.Q., and W.L. Silver, 2004, Some aspects of ecophysiological and biogeochemical responses of tropical forests to atmospheric change, *Philosophical Transactions of the Royal Society of London, Series B*, **359**, 463–476.

invasion of nonnative plants.[7] The lack of understanding of fundamental biogeochemical and ecological processes limits our ability to predict the ultimate consequences of elevated CO_2 on land carbon balance.

Background. Investigation of the effects of elevated CO_2 on land carbon balance has relied on manipulative experiments and natural gradient studies to isolate the physiological responses of plants on a variety of time scales. The CCSP strategic plan calls for augmentation of these manipulative studies, including addition of factors such as nitrogen or ozone, to improve the understanding of ecosystem response to climate change. Additional studies must be conducted in a variety of ecosystem types and include participation by a diverse range of scientists to study the physiological processes that mediate plant response to elevated CO_2, the mechanisms (and time scale) by which those changes in plant carbon balance are translated into ecosystem carbon storage, and the spatial variability of edaphic factors (e.g., climate, nutrient availability) that regulate the magnitude of response of individual plants and ecosystems. Improved understanding of these processes will ultimately be incorporated into models that estimate the magnitude of the global carbon land balance.

Adaptive Management of Water Resources

Related CCSP Questions, Milestones, and Products. Question 4.5: "How can information on climate variability and change be most efficiently developed, integrated with non-climatic knowledge, and communicated in order to best serve societal needs?"[8] Relevant milestones and products can be grouped into three themes: (1) develop experimental hydrologic forecasting and decision support systems that take advantage of emerging CCSP data and information; (2) pilot those systems in specific operational settings, using them in parallel with current forecasting and decision support systems; and (3) pilot use of new information in existing decision support systems.[9]

---

[7]Isebrands, J.G., E.P. McDonald, E. Kruger, G. Hendrey, K. Percy, K. Pregitzer, J. Sober, and D.F. Karnosky, 2001, Growth responses of *Populus tremuloides* clones to interacting elevated carbon dioxide and tropospheric ozone, *Environmental Pollution*, 115, 359–371; Krupa, S., 2003, Atmosphere and agriculture in the new millennium, *Environmental Pollution*, 126, 293–300.

[8]Climate Change Science Program and the Subcommittee on Global Change Research, 2003, *Strategic Plan for the U.S. Climate Change Science Program*, Washington, D.C., p. 50.

[9]Climate Change Science Program and the Subcommittee on Global Change Research, 2003, *Strategic Plan for the U.S. Climate Change Science Program*, Washington, D.C., pp. 59–62.

Rationale. Two goals in Chapter 11 ("Decision Support Resources Development") of the CCSP strategic plan are support for adaptive management (largely at the regional level) and support for operational decisions on climate variability and change. Developing information resources is central to each goal. Recent reports highlight the need for new information to support water resources management and other water-related decisions.[10] Explosive population growth and changing climate have combined to create imbalances between water supply and demand.

Background. Solutions to water shortages are limited by both inadequate institutional structures and inadequate physical facilities. As water becomes an increasingly valuable commodity, more accurate information will be needed to support estimates of the volume of natural water reservoirs (e.g., snow pack, groundwater), to understand fluxes (e.g., evapotranspiration, groundwater recharge), to perform hydrologic modeling (e.g., stream flow forecasting), and to support decision making. A program to collect such information could begin with the development of measurement networks, data management systems, and integrative tools (e.g., models) (1) to support water resources research and (2) to identify specific process studies and regions in which new information can both enhance existing decision support methods and encourage the use of emerging, experimental, decision support tools. In the long term, advances in research measurement networks, data systems, and integrative tools could bring new knowledge and technology into routine use in a variety of applications and operations. Research and interactions with decision makers are needed to address the efficacy of well-established practices and forecast methods that are based on historical data and performance. Such interaction is also necessary to determine information needs for future climate conditions that may lie outside the range of past system behavior. Thus, a sustained research effort that allows concurrent use of existing and emerging technology will be essential to demonstrate that the new tools will improve decision making.

CONCLUSIONS

Analysis of the case studies above and in Appendix B has revealed a number of challenges in applying and generalizing metrics.

---

[10]U.S. Bureau of Reclamation, 2003, Water 2025: Preventing Crises and Conflict in the West, <http://www.doi.gov/ water2025.pdf>.

TABLE 5.3 Adaptive Management of Water Resources

Type	Example Metrics
Process	• Does the CCSP have an effective planning structure, involving both agency managers and the scientific community, that is used to set priorities and implement water resource programs? • Does an adequate structure exist for peer review of both CCSP water resource programs and the research supported by those programs? • Does the CCSP support programs that effectively sustain research-applications partnerships, carry out a continuing assessment process, and provide test-beds for emerging water resource information and decision support tools? • Is the science in these planned programs responsive to the needs of regional stakeholders? • Does the CCSP water resources plan provide for the measurements, modeling, and decision support needed to link water cycle research and operational needs?

Input	• Annual R&D expenditures are sufficient to implement and sustain the following:
	  — Principal investigator (PI) and/or "centers" projects directed toward achieving the objectives
	  — Investigation of
	    ◦ Competing ideas and interpretations of causes
	    ◦ Competing interpretations of data
	    ◦ Innovative approaches for gathering or interpreting water resources data
	• Funds are available for the development and maintenance of a sustainable water resources scientific community of sufficient depth and diversity

Output	• Established (accepted, peer-reviewed, published) baselines for hydrologic forecasting improved as a result of CCSP-supported research • Consistent and reliable estimates and forecasts of water resources quantities (e.g., volume of natural water reservoirs, fluxes) to support adaptive management • Water resource planning scenarios that take into account contingencies such as substantial decreases in mountain snowpack expected as a result of further climate warming or multiyear droughts that stress water resources systems well beyond their design capacity • Accurate regional and national measures of the hydrologic effects likely associated with climate change • Quantitative information on components of the regional, national, and global water cycle that are important for water resources management, such as precipitation patterns and trends, streamflow trends, snowpack, and groundwater changes • Establishment of the degree to which these components are changing because of factors other than natural variability, such as moisture fluxes and precipitation • Sustainable information systems that make water resource data and information readily available to research and applications users

TABLE 5.3 Continued

Type	Example Metrics
Outcome	• Effective pilot research-applications partnerships result in experimental use of more accurate hydrologic forecasting tools and improved decision making • A regional demand exists among stakeholders for emerging CCSP data and information to support decision making • Decision support systems have been adapted to use emerging CCSP data and information • Improved information and technology have resulted in improved operational management of water resources, such as water allocations and reservoir operations • New infrastructure (e.g., groundwater backup systems for surface reservoirs) provides a more stable supply of water • More effective water resources planning structures, such as state drought task forces and agency capital investment plans, have been initiated that explicitly consider climate change
Impact	• Increased resilience of the water supply has decreased the vulnerability of populations to hydrologic aspects of climate variability and change

Challenges in the Application of Metrics

1. Several case studies revealed that a number of metrics are not amendable to numerical scores and require qualitative assessments. In general, the higher-order themes (e.g., theme 6—improving the assessment and management of risk) and higher-order measures (e.g., outcome and impact) appear to be more amenable to qualitative assessments based on peer review and stakeholder analysis. However, numerical scores may also be difficult to apply to lower-order themes such as data collection and analysis. For example, the design of a sampling scheme (necessary accuracy, precision, sampling scale) for climatic forcing factors and analysis of the results requires expert judgment (a) to avoid aliasing (the inevitable tendency of high-frequency components to appear to the observer as erroneous lower-frequency components or even space-time mean values if sampling criteria are not met) or (b) to recognize valuable uses of data, even when data accuracy proves to be less than the a priori measurement requirements. The case study of paleoclimate time series showed that no simple quantitative metric, such as the number of cores examined, would establish a milestone in understanding. Sometimes one core is sufficient to transform our views of ancient climate history, and additional observations may prove useful only for establishing generality. The goal should not be to increase the number of records but to improve their quality and usefulness for interpreting past climate. Achieving this goal requires expert judgment.

2. The CCSP plan is characterized by a large number of milestones and products, many of which are dependent on reaching other milestones. For example, metrics associated with the assessment and management of risk related to public health threats (theme 7) depend on scientific advances in a number of areas, including climate prediction, the linkages between environment and health, and the fundamental ecology of infectious diseases. Reaching these interdependent milestones could (a) involve different (and multiple) leaders, perhaps serving agencies with different missions and success criteria; (b) reflect different capabilities in observing and modeling; and (c) involve different assessments of the level of understanding, depending on whether the milestone involves social, biological, or atmospheric sciences. The difficulty of integrating all of these dependencies presents a significant challenge to producing and implementing useful metrics.

3. The case studies revealed concerns about whether a weak score on one element could dominate the evaluation of an otherwise strong program. Low scores generally indicate where work is needed, and high scores indicate substantial progress or successes. However, a high score does not mean that improvements cannot be made. Similarly, not every low score indicates problems. Low scores on outcome and impact measures may reveal that the science is still in the discovery phase, not that the program element is a failure. Moreover, a low score on lower-order metrics may not preclude a high score on higher-order metrics. For example, an output metric in theme 2 (improved accuracy of measuring sea level) might receive a low score, but the observations might inform a more important outcome metric (e.g., measurements of sufficient accuracy to inform assessments and policy) or lead to unanticipated outcomes. In many areas of interest, the success of a program can be evaluated only in the context of the sometimes myriad uses to which it can be put.

4. The evolution of knowledge can have a cascading effect on metrics. For example, theme 3 (understanding processes) measures frequently include an element of testing predictions against measurable quantities and periodic assessment of forecast ability. These, in turn, require involvement of the scientific community in activities designed to ensure that measurements made at many different locations and times are suitable for testing understanding of process models. Increased understanding may lead to new requirements that exceed the original data collection requirements. Additional progress may therefore depend on additional coordination of measurement networks and cooperation by the scientific community.

5. Because of the importance of "reducing uncertainties" within the CCSP strategic plan, numerous case studies examined potential metrics associated with complexity and uncertainty. Because complexity will remain or increase even as knowledge advances, scores for output and outcome metrics may not improve significantly with time. Similarly, the number of

uncertainties associated with climate change will likely be reduced through research and expanding knowledge. However, many of these will be replaced with new uncertainties, preventing metric scores from improving. Consequently, it is necessary to develop decision structures that can assess evolving uncertainties.

6. Impact metrics may prove difficult to quantify on a routine basis. First, the time scales associated with assessing impact often exceed the time frame for policy decisions. National policy making about a climate response also necessarily takes place in a political context. Thus, the salience of particular issues and the timeliness of different assessment methods and case analyses may depend on year-to-year national and international events. This circumstance places a premium on the development of a wide portfolio of analysis tools and applications. However, it is difficult to imagine an impact metric for performance of such a portfolio.

7. The importance of peer review, expert opinion, and stakeholder judgments was noted in the majority of case studies.

8. It may be a significant challenge to evaluate priorities between different measurement areas that involve the needs of different disciplines.

Ability to Generalize

Process and input metrics appear to be easily generalized among case study topics. Higher-order measures (output, outcome, and impact) appear to be more specific both to the tasks required to achieve the overall objective and to the unique features of the specific domains. As such, they do not appear to be directly transferable. However, even these measures can be generalized if the overall objectives are taken into consideration, such as improved knowledge of processes, improved forecasting capability, improved understanding of uncertainties and limitations, and improved management. The key to developing generalized metrics is the level of aggregation. For example, modeling needs for the assessment of broad-scale, long-term global environmental agreements are very different from those required for policies on specific technologies, such as pumping CO_2 into depleted oil fields. It may be possible, however, to develop process metrics to assess, at a higher level of aggregation, goals concerned with the establishment of a capacity to carry out policy-relevant integrated studies, the development of a portfolio of different capabilities, or the degree to which various studies are useful for policy making, without addressing a specific policy.

6

Metrics for the
Climate Change Science Program

$\mathcal{T}$he previous chapter shows how the committee developed metrics for specific Climate Change Science Program (CCSP) objectives. This chapter proposes a set of metrics to assess progress of any CCSP program element and guide future strategic planning.

DEVELOPMENT OF GENERAL METRICS

Comparison of all the example metrics created by the committee showed that process and input measures tend to be similar in all of the case studies, whereas output, outcome, and impact measures tend to be more specific to the case study goal. However, some of these output, outcome, and impact metrics could be rewritten more generically (see examples in Table 6.1). This observation raised the possibility that a single set of metrics with broad application to the CCSP could be devised.[1] Such a set would potentially be far more useful to the CCSP than a long list of highly specific metrics.

---

[1] A similar exercise examining environmental performance metrics in four industry sectors (automotive, chemical, electronics, and pulp and paper) yielded the same conclusion. A number of metrics were relatively common across the sectors and the use of general metrics made a significant contribution to the ability to assess competitive performance. See National Academy of Engineering and National Research Council, 1999, *Industrial Environmental Performance Metrics: Challenges and Opportunities*, National Academy Press, Washington, D.C., 252 pp.

TABLE 6.1 Examples of the Way Metrics Specific to Individual Case Studies Were Worded Generically

Case Study Wording	Generic Wording
Output Metrics	
• Development of a suite of new measurement techniques that are capable of detecting carbon allocation patterns on time scales of (1) hours, (2) days to weeks, and (3) a growing season in response to external variables and photosynthetic rates of plants in control versus experimentally manipulated systems	• The program results in peer-reviewed and broadly accessible results, such as (1) data and information, (2) new and applicable measurement techniques, (3) scenarios and decision support tools, and (4) well-described and demonstrated relationships that improve our understanding of processes or enable forecasting and prediction
• Production of a facility that (1) can be put into the field for years at a time and (2) can maintain atmospheric CO_2 levels at a specific set point (e.g., 50 ppm [parts per million] above ambient levels), with a precision (averaged over 1 hour) of 5 ppm • Sustainable information systems that make water resource data and information readily available to research and applications users	• Adequate community and/or infrastructure to support the program has been developed
Outcome Metrics	
• Are the aerosol measurements together with other aerosol research resulting in better understanding of the uncertainties in climate projections due to direct and indirect aerosol processes? • Are the research results leading to lower uncertainties in the historical contributions to sea-level rise and thence to better projections of future sea-level rise?	• The program has led to the identification of uncertainties, increased understanding of uncertainties, or reduced uncertainties
• Consistent and reliable projections of vegetation change and climate-vegetation interactions and feedbacks, with well-described sources of error and limitation • A peer reviewed, published, broadly accepted conclusion about our ability to simulate the twentieth century climate and attribute these variations to specific causes	• The program has yielded improved understanding, such as (1) quantification of important phenomena or processes, (2) more consistent and reliable predictions or forecasts, (3) increased confidence in our ability to simulate and predict climate change and variability, and (4) peer-reviewed, published, broadly accepted conclusions about key issues or relationships

TABLE 6.1 Continued

Case Study Wording	Generic Wording
• Ability to predict the extent to which a change in climate will significantly affect public health, as measured by an increase in infant mortality rates, declines in human life expectancy, or other factors • Consistent and reliable estimates and forecasts of water resources quantities (e.g., volume of natural water resources, fluxes) to support adaptive management • Technology developed for rapid control of trace gas concentrations at high precision	• The measurements, analysis, and results are being used (1) to answer the high-priority climate science questions that motivated them, (2) to address objectives outside the program plan, or (3) to support beneficial applications and decision making, such as forecasting, cost-benefit analysis, or improved assessment and management of risk

Impact Metrics

• Significantly reduced morbidity and mortality rates as a result of improved management of infectious disease • "No-build" zones established between new structures (e.g., roads, railways, houses) and the shoreline protect communities from sea-level rise	• The program has benefited society in terms of enhancing economic vitality, promoting environmental stewardship, protecting life and property, and reducing vulnerability to the impacts of climate change

The committee tested the concept by first combining the metrics in each category into a master list (Appendix B). The metrics within each category were checked for consistency with the definitions in Box 1.3 and examined for uniqueness, similarity, or overlap. Next, generic wording was developed for process, input, output, outcome, and impact measures, which required some rearranging and grouping. The metrics were written to permit a yes-no answer or a 1-5 score, although other scoring schemes (e.g., Army's red, yellow, green light approach)[2] could also be used.

The general metrics in Box 6.1 emerged from the iterative process described above. Note that the rankings will have to be defined for each measure, as exemplified in Chapter 2 (see example 1-5 ranking for metrics in Table 2.3). The type of ranking (e.g., yes or no, 1-5 scale, or some combination) is a matter of preference of the program leader.

---

[2]Department of Defense, 2003, Performance and Accountability Report: Fiscal Year 2003, Washington, D.C., p. 381, <http://www.defenselink.mil/comptroller/par/fy2003/00_Entire_Document.pdf>.

Box 6.1
General Metrics for the CCSP

Process Metrics (measure a course of action taken to achieve a goal)

1. Leader with sufficient authority to allocate resources, direct research effort, and facilitate progress.
2. A multiyear plan that includes goals, focused statement of task, implementation, discovery, applications, and integration.
3. A functioning peer review process in place involving all appropriate stakeholders, with (a) underlying processes and timetables, (b) assessment of progress toward achieving program goals, and (c) an ability to revisit the plan in light of new advances.
4. A strategy for setting priorities and allocating resources among different elements of the program (including those that cross agencies) and advancing promising avenues of research and applications.
5. Procedures in place that enable or facilitate the use or understanding of the results by others (e.g., scientists in other disciplines, operational users, decision makers) and promote partnerships.

Input Metrics (measure tangible quantities put into a process to achieve a goal)

1. Sufficient intellectual and technologic foundation to support the research.
2. Sufficient commitment of resources (i.e., people, infrastructure, financial) directed specifically to allow the planned program to be carried out.
3. Sufficient resources to implement and sustain each of the following: (a) research enabling unanticipated scientific discovery, (b) investigation of competing ideas and interpretations, and (c) development of innovative and comprehensive approaches.
4. Sufficient resources to promote the development and maintenance of each of the following: (a) human capital; (b) measurement systems, predictive models, and synthesis and interpretive activities; (c) transition to operational activities where warranted; and (d) services that enable the use of data and information by relevant stakeholders.
5. The program takes advantage of existing resources (e.g., U.S. and foreign historical data records, infrastructure).

Output Metrics (measure the products and services delivered)

1. The program produces peer-reviewed and broadly accessible results, such as (a) data and information, (b) quantification of important phenomena or processes,

Scientific programs yield a continuum of products and activities. Therefore, distinguishing between output and outcome measures and between outcome and impact measures requires some care. In this report, output metrics are tangible products and services, including scientific results and new techniques, capabilities, or infrastructure. Outcome metrics are broader

(c) new and applicable measurement techniques, (d) scenarios and decision support tools, and (e) well-described and demonstrated relationships aimed at improving understanding of processes or enabling forecasting and prediction.

2. An adequate community and/or infrastructure to support the program has been developed.

3. Appropriate stakeholders judge these results to be sufficient to address scientific questions and/or to inform management and policy decisions.

4. Synthesis and assessment products are created that incorporate these new developments.

5. Research results are communicated to an appropriate range of stakeholders.

Outcome Metrics (measure results that stem from use of the outputs and influence stakeholders outside the program)

1. The research has engendered significant new avenues of discovery.

2. The program has led to the identification of uncertainties, increased understanding of uncertainties, or reduced uncertainties that support decision making or facilitate advance of other areas of science.

3. The program has yielded improved understanding, such as (a) more consistent and reliable predictions or forecasts, (b) increased confidence in our ability to simulate and predict climate change and variability, and (c) broadly accepted conclusions about key issues or relationships.

4. Research results have been transitioned to operational use.

5. Institutions and human capacity have been created that can better address a range of related problems and issues.

6. The measurements, analysis, and results are being used (a) to answer the high-priority climate questions that motivated them, (b) to address objectives outside the program plan, or (c) to support beneficial applications and decision making, such as forecasting, cost-benefit analysis, or improved assessment and management of risk.

Impact Metrics (measure the long-term societal, economic, or environmental consequences of an outcome)

1. The results of the program have informed policy and improved decision making.

2. The program has benefited society in terms of enhancing economic vitality, promoting environmental stewardship, protecting life and property, and reducing vulnerability to the impacts of climate change.

3. Public understanding of climate issues has increased.

results, such as improved scientific understanding or reliable forecasts, that influence stakeholders outside the program, including scientists working in other fields, resource managers, and policy makers. The level of influence of impact metrics is even greater and includes the long-term results of actions by managers, policy makers, and science and business leaders.

ROBUSTNESS OF THE GENERAL METRICS

The applicability of the general set of metrics was tested by applying them to four CCSP program elements or related programs at different scales:

1.   CCSP Question 4.1: To what extent can uncertainties in model projections due to climate system feedbacks be reduced?[3]
2.   Assessment 2.4: Assessment of trends in emissions of ozone-depleting substances, ozone layer recovery, and implications for ultraviolet radiation exposure and climate change.[4]
3.   Chapter 11, Goal 2-related issue: Develop resources to support adaptive management and planning for responding to climate variability and climate change, and transition these resources from research to operational application.[5]
4.   CCSP and U.S. Global Change Research Program (USGCRP) in its entirety.[6]

The test cases were not intended to actually assess progress in four particular program elements. Such an assessment is beyond the committee's charge and capability and is left to CCSP program managers and appropriate stakeholder groups. Consequently, the committee's answers, scores, and associated explanation of the metrics of the test cases are not provided. However, some insights into the application of the general metrics to the four test cases are given below.

For each test case, the committee evaluated the use of the metrics by actually answering or scoring each question. In some cases, especially those involving complex or qualitative measures, significant explanation describing the progress and performance had to accompany the answer or ranking. These tests led to some iteration and improvement in the wording, but the committee concluded that the general set of metrics is robust and could be used to measure progress and guide strategic thinking across the entire CCSP.

---

[3]Climate Change Science Program and Subcommittee on Global Change Research, 2003, *Strategic Plan for the U.S. Climate Change Science Program*, Washington, D.C., pp. 42–44.
[4]Climate Change Science Program and Subcommittee on Global Change Research, 2003, *Strategic Plan for the U.S. Climate Change Science Program*, Washington, D.C., p. 27.
[5]Climate Change Science Program and Subcommittee on Global Change Research, 2003, *Strategic Plan for the U.S. Climate Change Science Program*, Washington, D.C., p. 114.
[6]Climate Change Science Program and Subcommittee on Global Change Research, 2003, *Strategic Plan for the U.S. Climate Change Science Program*, Washington, D.C., 202 pp.; Climate Change Science Program and Subcommittee on Global Change Research, 2002, *Our Changing Planet: The Fiscal Year 2003 U.S. Global Change Research Program and Climate Change Research Initiative*, Washington, D.C., 124 pp.

Test Case 1: CCSP Question 4.1 (To what extent can uncertainties in model projections due to climate system feedbacks be reduced?). In this test case, the process metrics generally received low scores because there is little focused planning and leadership is spread across different modeling programs. Although valuable collaborations exist at the project level and through international programs, their purpose is to better understand the model results, rather than to direct modeling efforts. However, a number of the output and outcome metrics received high scores because the models are improving and are leading to improved understanding of the climate system. Many of the outcome metrics and all of the impact metrics were difficult to score because they require qualitative judgments, such as peer review or stakeholder assessment. Moreover, it may take decades to assess the impact of model improvements.

The fact that good scientific outcomes are sometimes possible without extensive institutional planning or focused leadership is not surprising. Scientists are generally capable of identifying research approaches and developing grass roots collaborations without formal direction, as long as the agencies maintain an environment that promotes discovery and innovation. However, strategic planning can speed scientific outcomes, as illustrated in the second test case below.

Test Case 2: Assessment 2.4 (Assessment of trends in emissions of ozone-depleting substances, ozone layer recovery, and implications for ultraviolet radiation exposure and climate change). The committee used the history of stratospheric ozone depletion from the mid-1970s to the mid-1980s as summarized in Chapter 2 to evaluate this test case. Once the stratospheric ozone program emerged from the discovery phase in the mid-1970s, it led to numerous scientific and policy successes. The high scores given retrospectively to metrics across the board are in agreement with this perception. This test case suggests that the combination of a strong plan, an active research community enabled to make scientific discoveries, and leadership committed to using the scientific output leads to a highly successful program.

Test Case 3: Chapter 11, Goal 2-Related Issue (Develop resources to support adaptive management and planning for responding to climate variability and climate change, and transition these resources from research to operational application). The National Oceanic and Atmospheric Administration's (NOAA's) Regional Integrated Science and Assessments (RISA) program[7] was used to evaluate this test case. The RISA

---

[7]See <http://www.ogp.noaa.gov/mpe/csi/risa/>.

program supports research on climate-sensitive issues of concern to decision makers and policy planners at a regional level. The research, which is largely carried out by seven university-government-private sector consortia, focuses on fisheries, water, wildfire, agriculture, public health, and coastal restoration. Examination of the RISA program revealed the presence of a plan and appropriate leadership, but few peer-reviewed results and limited funds to promote discovery and innovation. Although the RISA program enables significant output related to stakeholder needs, weaknesses in process (peer review) and input (sufficient support) limit the ultimate outcomes.

The committee found it difficult to score a number of the metrics in this test case because the information needed to make the evaluation (e.g., use of results outside the program) has not been collected. This will likely be the case for the first few evaluations of any program. Experience will show which metrics are most important and what information is needed to evaluate them regularly.

Test Case 4: CCSP and USGCRP in Its Entirety. The analysis of the CCSP-USGCRP as a program revealed a different set of issues. Many of the process metrics reveal weaknesses. Although the CCSP has central leadership, the day-to-day leadership of its many programs and activities is distributed among different agencies. Distributed leadership also affects many of the factors that are related to other process metrics, such as priority setting and establishment of peer review systems. The success of the applied parts of the program in particular may well fall short without better coordination and leadership.

Scores on the input metrics were mixed. The comprehensive nature of CCSP goals ensures that many aspects of the program will be resource limited. However, although funding is insufficient to accomplish everything in the plan, it allows both unfettered and mission-oriented research. As a result, the program has produced significant outputs and outcomes.

Assessing the impacts of such a far-reaching and complex program presents a considerable challenge. First, impacts depend on a number of factors (e.g., politics, technological advances), many of which are not connected to the CCSP. Only a fraction of the scientific outcomes may have significant impact on policy and decision making, and those outcomes will themselves depend on the success of many other program elements. Second, it may take decades to assess the impact of the CCSP and its predecessor USGCRP. The two- to four-year time frame of the CCSP milestones, products, and payoffs limits the number of impacts that the CCSP can claim. Nevertheless, it is clear that the CCSP-USGCRP has made substantial contributions to the global debate on climate change. With the perspective of time, the magnitude of this impact will become more clear and is likely to grow substantially.

USE OF GENERAL METRICS TO SET PRIORITIES

One of the more difficult problems for agency managers and Office of Management and Budget (OMB) budget examiners to address is setting priorities among different types of programs in the absence of an overarching national strategy on environmental science issues. The general metrics may provide a useful starting point for choosing between different projects. They could be applied to each project and the results (scores plus commentary) compared. The comparison is simplest when similar program elements are being considered, such as land and ocean observing programs. In such cases, the factors needed to measure process, inputs, and outputs are similar and the comparison is straightforward. However, even when the goals (and thus the process, input, and outputs) are different, the general metrics facilitate identification of the strengths and weaknesses of different programs, including the readiness of a program to advance beyond the discovery stage or the effect of resource limitations on particular parts of the program. Insights gained from such a comparison could provide the basis for a more informed discussion of priorities than currently exists.

CONCLUSIONS

Overall, the committee found that the general metrics listed in Box 6.1 provide a useful starting point for CCSP program managers to assess program performance and identify barriers to progress. Tests performed by the committee suggest that the general metrics are also likely to be applicable to any science program that has established goals. The list appears to work for programs at all levels of granularity, although not all metrics will apply to all programs. In addition to providing a yes or no answer or a numerical score, the formal evaluation should include a commentary explaining the meaning of the score. Indeed, an explanation of the meaning of the measure is required in Program Assessment Rating Tool (PART) reports.[8] This commentary is as important as the specific answer or score.

---

[8]Office of Management and Budget, 2005, Guidance for Completing the Program Assessment Rating Tool (PART), pp. 13–14, <http://www.whitehouse.gov/omb/part/fy2005/2005_guidance.doc>.

7

Conclusions and Next Steps

*T*he committee was charged to identify quantitative performance measures and metrics to assess progress in three to five areas of climate change research. The committee began by selecting a representative set of Climate Change Science Program (CCSP) objectives and developing a long list of metrics for each. However, analysis of the measures specific to these objectives showed that a general set of metrics could be developed and used to assess the progress of any element of the CCSP (Chapter 6). This unexpected conclusion, combined with the principles (Chapter 3), led the committee to think about metrics not just as simple ways to gauge progress, but as a tool to guide strategic planning and to foster future progress. The committee believes that the general metrics have the potential to be far more useful to CCSP agencies than a few specific metrics in selected areas of climate change research. The answers to the charge given below are presented in the context of this major conclusion.

ANSWERS TO THE COMMITTEE CHARGE

1. *Provide a general assessment of how well the CCSP objectives lend themselves to quantitative metrics.*

Meaningful metrics can be developed for all of the CCSP objectives. Some of the metrics will be quantitative, especially those that measure inputs (e.g., amount of resources devoted to the program) and outputs (e.g., creation of new products or techniques). However, most will be qualitative, especially those that focus on the research and development process, the outcome of research, and its impact on society. These generally require peer review (e.g., to evaluate scientific quality) or stakeholder assessments.

CCSP objectives range from the general (e.g., overarching goals) to the specific (e.g., milestones, products, and payoffs). The more general the objective, the greater is the number of qualitative contributing factors and the less quantitative are the metrics. For example, improvements to databases of water cycle variables can generally be measured quantitatively, but the resulting improvement in drought prediction models and resource decisions that use those predictions will require increasingly subjective analysis and a greater emphasis on expert assessment.

2. *Identify three to five areas of climate change and global change research that can and should be evaluated through quantitative performance measures.*

Both the Office of Management and Budget (OMB) and the agencies participating in the CCSP are seeking a manageable number of quantitative performance measures to monitor the progress of the program. The metric cited most often is the reduction of uncertainty (Chapter 4). However, by itself, reduction of uncertainty is a poor metric because (1) uncertainty about future climate states may increase, decrease, or remain the same as more is understood about the governing elements, and (2) the data needed to calculate errors in the probability estimates are limited or nonexistent. The danger of using this metric is that increasing uncertainty might be interpreted as a failure of the program, when the reverse may well be true.

The committee agrees that a limited set of metrics should be chosen. It would be expensive to implement all possible measures, and the results may be difficult for individual agencies to use to manage their programs and demonstrate success to Congress, OMB, and the public. However, the CCSP strategic plan provides neither a sense of priorities nor a definition of success.[1] Indeed, a National Research Council review of the CCSP strategic plan noted that "many of the objectives in the plan are too vaguely worded

---

[1]The CCSP strategic plan does not list measures of success, but examples can be found elsewhere. For instance, measures of success proposed for the Department of Energy's program on environmental quality research include the degree to which the program has led to improved performance, reduced risks to human health or the environment, decreased costs, and advanced schedules. See National Research Council, 2001, *A Strategic Vision for Department of Energy Environmental Quality Research and Development*, National Academy Press, Washington, D.C., 170 pp.

to determine what will constitute success."[2] Such guidance is essential for narrowing down research areas for which metrics should be developed.

However, even if such guidance were available, focusing on metrics in a few areas of climate change and global change research might not be useful for managing the program and achieving successful outcomes. The key to promoting successful outcomes is to consider the program from end to end, starting with program processes and inputs and extending to outputs, outcomes, and long-term impacts. The general metrics developed by the committee (Box 6.1) provide a starting point for making this evaluation.

3. *For these areas, recommend specific metrics for documenting progress, measuring future performance (such as skill scores, correspondence across models, correspondence with observations), and communicating levels of performance.*

The list of general metrics can be used for any element of the CCSP. Quantitative measures in that list include the following:

- A multiyear plan that includes goals, focused statement of task, implementation, discovery, applications, and integration.
- Sufficient commitment of resources (i.e., people, infrastructure, financial) directed specifically to allow the planned program to be carried out.
- Synthesis and assessment products are created that incorporate new developments.

Generic quantitative metrics developed elsewhere for research and development (Appendix C) are also applicable to CCSP research elements. However, although these measures are useful for management, few scientific programs would wish to be judged on such terms. A mixture of qualitative and quantitative metrics would better capture the scope of CCSP objectives. A similar conclusion has been reached about measuring progress in other science programs.[3]

---

[2]National Research Council, 2004, *Implementing Climate and Global Change Research: A Review of the Final U.S. Climate Change Science Program Strategic Plan*, The National Academies Press, Washington, D.C., p. 26.

[3]National Research Council, 1996, *World-Class Research and Development: Characteristics for an Army Research, Development, and Engineering Organization*, National Academy Press, Washington, D.C., 72 pp.; National Science and Technology Council, 1996, Assessing Fundamental Science, <http://www.nsf.gov/sbe/srs/ostp/ assess/start.htm>; Cozzens, S.E., 1997, The knowledge pool: Measurement challenges in evaluating fundamental research programs, *Evaluation and Program Planning*, 20, 77–89; National Research Council, 1999, *Evaluating Federal Research Programs: Research and the Government Performance and Results Act,*

Although worded generically, the metrics listed in Table 6.1 can be rephrased to be specific to the program element being evaluated. This task is best carried out by agency managers because they have more complete knowledge of the program than any outside group could have. Moreover, the process of refining the metrics will be as valuable to the agencies as the measures themselves. A process for narrowing down and rephrasing the committee's list of general metrics is described in the following section.

4. *Discuss possible limitations of quantitative performance measures for other areas of climate change and global change research.*

Quantitative metrics can be developed for any CCSP objective. However, because quantitative metrics primarily (and only partially) measure inputs and outputs, they tell only a fraction of the story. The outcomes and impacts, which are the program results most visible to the public and usable to decision makers, are much more likely to be qualitative.

It may take years or even decades to assess the impact of CCSP programs, even though many are scheduled to produce results within two to four years. Answers to impact metrics will reflect the maturity of the programs as well as the complexity of the problems being analyzed. Consequently, many impact metrics developed for the CCSP will serve as a reminder of program goals, rather than a litmus test of achievement. Importantly, the CCSP will, with time, yield many unanticipated benefits because it supports discovery and innovation. General metrics that support successful outcomes, scenario planning, and other strategic improvements are more likely to reveal these unanticipated benefits than tightly specified, short-term objectives. A variety of such successful outcomes and impacts have already emerged from climate change programs that operated under the U.S. Global Change Research Program (USGCRP).

IMPLEMENTATION

Lessons from industry, academia, and federal agencies suggest that metrics are best used to support actions that allow programs to evolve

National Academy Press, Washington, D.C., 80 pp.; Geisler, E., 2000, *The Metrics of Science and Technology*, Quorum Books, Westport, Conn., 380 pp.; National Research Council, 2001, *Implementing the Government Performance and Results Act for Research: A Status Report*, National Academy Press, Washington, D.C., 190 pp.; National Research Council, 2003, *The Measure of STAR: Review of the U.S. Environmental Protection Agency's Science to Achieve Results (STAR) Research Grants Program*, The National Academies Press, Washington, D.C., 176 pp.

toward successful outcomes. Implementation of metrics should therefore be strategic and evolutionary, rather than fixed and prescriptive.

This report provides a guide to best practices (principles in Chapter 3) and a list of general metrics (Box 6.1) that the CCSP can use to evolve toward successful outcomes. The principles define prerequisites for assessing and enabling this evolution (e.g., leadership), as well as characteristics of useful metrics. The general metrics provide a way to think strategically about the program. Together, the principles and general metrics provide a framework for considering specific implementation issues. These range from how to evaluate the program, to ensuring that the metrics are reliable and valid, to factoring in the cost of evaluating and adjusting the measures on a regular basis.

Using the General Metrics

The way in which the general metrics are used depends on both on the identify of the evaluators and the granularity of the program or program elements to be evaluated. For example, an agency manager might quickly provide rough answers or scores to all of the general metrics for his or her program, identifying leaders, availability of resources, opportunities for innovation, the functioning of the peer review process, evidence of new techniques, number of peer-reviewed publications, and so forth. This assessment would allow the manager to assess strengths and weaknesses of the program and then determine an appropriate course of action.

Expert panels might use the general metrics as a framework for deeper exploration of issues, such as the value of different measurement techniques, assessment of the capabilities of new models, or the results of process studies. For example, output metric 1 (the program produces peer-reviewed and broadly accessible results, such as new and applicable measurement techniques) might prompt a thorough analysis of the types of new measurement techniques that were developed, as well as their veracity, limitations, and acceptance as a tool by the broader community. Expert panels may view process and input metrics only in hindsight, for example, tracing gaps in knowledge to limitations in program planning. Finally, an evaluation by stakeholders would likely emphasize outcomes and impacts. Depending on the nature of the program, this evaluation could be highly technical (e.g., the performance of a new flood forecasting model) or highly subjective (e.g., whether global change research has had noticeable impact on international policies).

Individual programs could easily spawn a set of quite specific metrics. For example, if the program plan calls for a doubling of the density of ocean buoys in the tropical Pacific, a metric might be the fraction of the new system that has been deployed. This program-dependent measure falls within the scope of input metric 4 (Box 6.1). Higher-level indicators (e.g., the degree to which the buoy array has improved seasonal to interannual forecasting or has outcomes that are recognized and utilized by different stakeholders) prompt equally specific output and outcome metrics and alternative modes of evaluation (e.g., expert review). The general metrics in Box 6.1 provide the categories to be evaluated, but they will have to be narrowed down and reworded in terms that are specific to the objective or program plan being evaluated (see examples in Table 6.1). In this manner, the general metrics serve as a template for evaluating programs and promoting progress.

It is important to note that the development of an optimum set of metrics is an iterative process. No one gets it right the first time. However, the process itself will yield valuable information about the program and how to continuously improve it.

Refining the Metrics

Once the key metrics have been identified, they must be refined to ensure that biases are recognized and minimized. An evaluation system must also be developed. An overview of these issues is given below.

Bias, Reliability, and Repeatability

Any measure that contains subjective factors or relies on judgments introduces estimation errors, biases, or inaccurate perceptions.[4] Yet subjective judgments are essential to evaluate both the scientific program elements of the CCSP and the usefulness of the resulting knowledge to users. Peer review, normally used to evaluate science quality, is subject to bias (e.g., against those who challenge conventional wisdom) and may not yield the same results from year to year.[5] User evaluations, often used to gauge the importance of knowledge that results from a program, are biased toward high satisfaction with free services. Having an appropriate mix of expertise on the evaluation team will minimize the chances of different groups obtaining different results and thereby increase the reliability and repeatability of the results.[6] The reliability of subjective measures could also be increased

---

[4]Werner, B.M., and W.E. Souder, 1997, Measuring R&D performance—State of the art, *Research Technology Management*, **March-April**, 34–42.

[5]Cozzens, S.E., 1997, The knowledge pool: Measurement challenges in evaluating fundamental research programs, *Evaluation and Program Planning*, 20, 77–89.

[6]Kostoff, R.N., 1998, Metrics for planning and evaluating science and technology, *R&D Enterprise—Asia Pacific*, 1, 30–33.

by aggregating multiple judgments or requiring different assessment teams to arrive at a consensus.[7]

Quantitative measures do not suffer from these limitations, although their objective nature can lend a false sense of credibility and validity.[8] They also overlook time lags that might bias the measurements. For example, research and development departments that have the same profit-to-expenditure ratios might produce results on different time horizons.

Finally, the selection of metrics themselves introduces biases and may also influence behaviors.[9] The values of the decision maker or evaluator are often mirrored in the selection and weighting of the measures. Also, once the metrics are known, they can bias behaviors to meet expectations built into the measures. The metric of number of papers published, for example, may lead scientists to publish a greater number of articles on the same research results. Being cognizant of these issues can sometimes minimize the influence of bias on metrics.

Aggregating Qualitative and Quantitative Measures

A suite of different kinds of metrics has been shown to be effective for science and technology programs. Such measures can be aggregated and compared using a variety of techniques. Formulas can be developed in which each class of measures is subjectively assigned a different weight. In the OMB Program Assessment and Rating Tool (PART) analysis, for example, the major classes of measures are weighted as follows:

- program purpose and design: 20 percent
- strategic planning: 10 percent
- program management: 20 percent
- program results and accountability: 50 percent (Appendix A, Box A.2)

Individual metrics can also be aggregated into more comprehensive measures that may include both quantitative and qualitative elements.[10] The different measures are assigned a weight and the qualitative measures are converted to numbers (e.g., an ordinal score or 0-100 percent of the

---

[7]Werner, B.M., and W.E. Souder, 1997, Measuring R&D performance—State of the art, *Research Technology Management*, **March-April**, 34–42.

[8]Werner, B.M., and W.E. Souder, 1997, Measuring R&D performance—State of the art, *Research Technology Management*, **March-April**, 34–42.

[9]Geisler, E., 2000, *The Metrics of Science and Technology*, Quorum Books, Westport, Conn., 380 pp.

[10]Werner, B.M., and W.E. Souder, 1997, Measuring R&D performance—State of the art, *Research Technology Management*, **March-April**, 34–42.

reference value). Care must be taken, however, to aggregate only measures that are well correlated.[11]

Finally, it is important to remember that any number describing a research activity or application depicts it imperfectly. Thus, agencies should not rely exclusively on the score of a particular metric or suite of metrics. The context, definition of scores, and commentary are at least as important as the specific answer or score and should be included in the formal evaluation.

Cost of Evaluating Metrics

The cost of developing and evaluating metrics must be balanced against the needs and resources of the program. Time costs can be considerable to develop an effective combination of quantitative and qualitative measures and to adjust them as experience reveals which are the most useful.[12] Professional training may even be required to develop qualitative measures that have validity and reliability. Collecting information to evaluate the metrics and normalizing and interpreting the data often take significant time, although time costs decline with subsequent evaluations. The highest time costs are for peer review evaluations.[13] Rather than peer review for every component of the CCSP, such investments should be targeted to improve management and performance of key program elements. All of these costs must be factored into determinations of how often the program should be evaluated to capture its impact over time.

The committee believes that a system of metrics, developed through an iterative process and evaluated in consultation with stakeholders, could be a valuable tool for managing the CCSP and further increasing its usefulness to society. For these metrics to be of real value, they must be implemented in a constructive fashion, following the guiding principles outlined in this report. That will require a great deal of thought by individual CCSP agencies as well as by the CCSP as a whole. Then, it will take time to determine whether these metrics help create a stronger and more successful CCSP. Thus, this report should be viewed as the first step and not as an end.

---

[11]Geisler, E., 2000, *The Metrics of Science and Technology*, Quorum Books, Westport, Conn., 380 pp.

[12]Werner, B.M., and W.E. Souder, 1997, Measuring R&D performance—State of the art, *Research Technology Management*, **March-April**, 34–42.

[13]Cozzens, S.E., 1997, The knowledge pool: Measurement challenges in evaluating fundamental research programs, *Evaluation and Program Planning*, 20, 77–89; Kostoff, R.N., 1998, Metrics for planning and evaluating science and technology, *R&D Enterprise—Asia Pacific*, 1, 30–33.

Appendixes

Appendix A

Measuring Government Performance

A number of federal laws and policies require government agencies to measure and report the performance of their programs. These include the Government Performance and Results Act (GPRA), the research and development (R&D) investment criteria, and the Program Assessment Rating Tool (PART). GPRA establishes a broad statutory framework for management and accountability, whereas the R&D investment criteria and PART are focused on more simplified measures of performance for budget decisions.

GOVERNMENT PERFORMANCE AND RESULTS ACT

The Government Performance and Results Act of 1993[1] was intended to increase the effectiveness, efficiency, and accountability of the federal government. It requires federal agencies to set strategic goals and to measure program performance against those goals. Reporting takes three forms:

1. a strategic plan, which states the agency mission, goals and objectives, and a description of how the goals and objectives will be achieved over the next five or more years;

2. an annual performance plan, which establishes performance goals as well as performance indicators for measuring or assessing the outputs, service levels, and outcomes of each program activity; and

---

[1]Public Law 103-62.

3. an annual performance report, which compares actual accomplishments with the performance goals.

The GPRA does not apply to interagency programs such as the Climate Change Science Program (CCSP). However, agency contributions to such programs are subject to GPRA, although they may be formulated in agency terms, rather than interagency terms.

Agencies that cannot express performance goals in an objective, quantifiable, and measurable form can seek Office of Management and Budget (OMB) approval for alternative forms. Science agencies have generally adopted both quantitative (e.g., publication count) and qualitative (e.g., progress in understanding) indicators.[2]

RESEARCH AND DEVELOPMENT INVESTMENT CRITERIA AND THE PROGRAM ASSESSMENT RATING TOOL

In 2002, two White House management initiatives—the R&D investment criteria and PART—were introduced in part to inform budget decisions. The R&D investment criteria were intended to improve the process for budgeting, selecting, and managing research and development programs.[3] Managers must demonstrate the extent to which their programs meet the tests of relevance, quality, and performance (see Box A.1). The criteria also address retrospective review of whether investments were well directed, efficient, and productive.

PART focuses on the subset of long-term and annual performance measures that capture the most important aspects of the program's mission and priorities. Based on a set of yes or no questions (see Box A.2), each program is assigned a score, which is translated into a qualitative rating: effective, moderately effective, adequate, ineffective, or results not demonstrated. The rating is intended to be used to tie program performance to

---

[2]General Accounting Office, 1997, *Measuring Performance: Strengths and Limitations of Research Indicators*, GAO/RCED-97-91, Washington, D.C., 34 pp.

[3]Memorandum on FY 2004 interagency research and development priorities, from John H. Marburger III, director of the Office of Science and Technology Policy, and Mitchell Daniels, director of the Office of Management and Budget, on May 30, 2002, <http://www.ostp.gov/html/ombguidmemo.pdf>. The guidelines drew heavily from National Research Council, 2001, *Implementing the Government Performance and Results Act for Research: A Status Report*, National Academy Press, Washington, D.C., 190 pp. OMB also developed guidelines for applied research, using the Department of Energy's (DOE's) applied energy technology programs as a pilot. See <http://www7.nationalacademies.org/gpra/Applied_Research.html>.

Box A.1
R&D Investment Criteria

The following criteria apply to all federal research and development programs.

Relevance

- Programs must have complete plans, with clear goals and priorities.
- Programs must articulate the potential public benefits of the program.
- Programs must document their relevance to specific presidential priorities to receive special consideration.
- Program relevance to the needs of the nation, of fields of science and technology, and of program "customers" must be assessed through prospective external review.
- Program relevance to the needs of the nation, of fields of science and technology, and of program "customers" must be assessed periodically through retrospective external review.

Quality

- Programs allocating funds through means other than a competitive, merit-based process must justify funding methods and document how quality is maintained.
- Program quality must be assessed periodically through retrospective expert review.

Performance

- Programs may be required to track and report relevant program inputs annually.
- Programs must define appropriate output and outcome measures, schedules, and decision points.
- Program performance must be retrospectively documented annually.

SOURCE: Office of Management and Budget, 2003, Budget Procedures Memorandum No. 861, Completing the Program Assessment Rating Tool (PART) for the FY 2005 Review Process, 60 pp., <http://www.whitehouse.gov/omb/part/bpm861.pdf>.

Box A.2
PART Questions and Relation to the R&D Investment Criteria

Program Purpose and Design (20 percent weighting)

Questions address R&D investment criteria of program relevance:

1.1 Is the program purpose clear?
1.2 Does the program address a specific and existing problem, interest, or need?
1.3 Is the program designed so that it is not redundant or duplicative of any other federal, state, local or private effort?
1.4 Is the program design free of major flaws that would limit the program's effectiveness or efficiency?
1.5 Is the program design effectively targeted so that resources will address the program's purpose directly and will reach intended beneficiaries?

Strategic Planning (10 percent weighting)

Questions address prospective aspects of the R&D investment criteria:

2.1 Does the program have a limited number of specific long-term performance measures that focus on outcomes and meaningfully reflect the purpose of the program?
2.2 Does the program have ambitious targets and time frames for its long-term measures?
2.3 Does the program have a limited number of specific annual performance measures that can demonstrate progress toward achieving the program's long-term goals?
2.4 Does the program have baselines and ambitious targets for its annual measures?
2.5 Do all partners (including grantees, subgrantees, contractors, cost-sharing partners, and other government partners) commit to and work toward the annual and/or long-term goals of the program?
2.6 Are independent evaluations of sufficient scope and quality conducted on a regular basis or as needed to support program improvements and evaluate effectiveness and relevance to the problem, interest, or need?
2.7 Are budget requests explicitly tied to accomplishment of the annual and long-term performance goals, and are the resource needs presented in a complete and transparent manner in the program's budget?
2.8 Has the program taken meaningful steps to correct its strategic planning deficiencies?

Additional questions for R&D programs:

2.RD1 If applicable, does the program assess and compare the potential benefits of efforts within the program and (if relevant) to other efforts in other programs that have similar goals?
2.RD2 Does the program use a prioritization process to guide budget requests and funding decisions?

Program Management (20 percent weighting)

Questions address prospective aspects of program quality and performance in the R&D investment criteria, as well as general program management issues:

3.1 Does the agency regularly collect timely and credible performance information, including information from key program partners, and use it to manage the program and improve performance?

3.2 Are federal managers and program partners (including grantees, subgrantees, contractors, cost-sharing partners, and other government partners) held accountable for cost, schedule, and performance results?

3.3 Are funds (federal and partners') obligated in a timely manner and spent for the intended purpose?

3.4 Does the program have procedures (e.g., competitive sourcing or cost comparisons, information technology improvements, appropriate incentives) to measure and achieve efficiencies and cost-effectiveness in program execution?

3.5 Does the program collaborate and coordinate effectively with related programs?

3.6 Does the program use strong financial management practices?

3.7 Has the program taken meaningful steps to address its management deficiencies?

Additional question for R&D programs:

3.RD1 For R&D programs other than competitive grants programs, does the program allocate funds and use management processes that maintain program quality?

Program Results and Accountability (50 percent weighting)

Questions address retrospective aspects of the R&D investment criteria, with emphasis on performance:

4.1 Has the program demonstrated adequate progress in achieving its long-term performance goals?

4.2 Does the program (including program partners) achieve its annual performance goals?

4.3 Does the program demonstrate improved efficiencies or cost-effectiveness in achieving program goals each year?

4.4 Does the performance of this program compare favorably to other programs, including government, private, etc., with similar purpose and goals?

4.5 Do independent evaluations of sufficient scope and quality indicate that the program is effective and achieving results?

SOURCE: Office of Management and Budget, 2005, Guidance for Completing the Program Assessment Rating Tool (PART), 64 pp., <http://www.whitehouse.gov/omb/part/fy2005/2005_guidance.doc>.

budget appropriations.[4] Twenty percent of federal programs are being rated each year, beginning with the fiscal year (FY) 2004 budget request.[5]

A 2004 General Accounting Office (GAO) report found that PART had helped structure OMB's use of performance information for program analysis and internal review.[6] However, budget allocations were not always tied to program ratings. Programs rated as "effective" or "moderately effective" did not always receive increased funding, and programs rated as "ineffective" did not always lose funding. The report also noted that by using PART to influence GPRA measures, OMB is influencing agency program goals, to the detriment of a wide range of stakeholders. It concluded that although PART is useful for program-level budget analysis, it cannot substitute for GPRA's longer-term, strategic focus on thematic goals. Nevertheless, goals and performance measures relevant to the R&D criteria and PART are being incorporated into future GPRA agency performance plans.[7]

---

[4]Office of Management and Budget, 2003, Performance Measurement Challenges and Strategies, 13 pp., <http://www.whitehouse.gov/omb/part/challenges_strategies.pdf>.

[5]Agency programs relevant to climate change that were evaluated in the FY 2004 budget include Department of Defense (DOD) Basic Research; DOE's Basic Energy Sciences, Biological and Environmental Research, Environmental Management, and Office of Science; U.S. Agency for International Development (USAID) Climate Change; and National Science Foundation (NSF) Geosciences. In FY 2005, relevant agency programs include the U.S. Department of Agriculture's (USDA's) National Resources Inventory and Soil Survey; Department of Interior's (DOI's) Science and Technology; Environmental Protection Agency's (EPA's) Ecological Research; and National Aeronautics and Space Administration's (NASA's) Biological Sciences Research and Earth Science Applications. See <http://www.whitehouse.gov/omb/part/program_assessments_planned_2005.html>.

[6]General Accounting Office, 2004, *Performance Budgeting: Observations on the Use of OMB's Program Assessment Rating Tool for the Fiscal Year 2004 Budget*, GAO-04-174, Washington, D.C., 67 pp.

[7]Office of Management and Budget, 2003, Budget procedures memorandum no. 861, Completing the Program Assessment Rating Tool (PART) for the FY2005 review process, 60 pp., <http://www.whitehouse.gov/omb/part/ bpm861.pdf>.

Appendix B

Case Study Metrics for the Climate Change Science Program

This appendix provides case study examples of metrics for a range of program elements drawn from the Climate Change Science Program (CCSP). Each case study is focused on specific CCSP questions and milestones and includes a rationale and background information needed to inform the development of metrics. Specific process, input, output, outcome, and impact metrics developed by the committee appear at the end of each case study (Tables B.1-B.8). Following the case studies the metrics are grouped together (Tables B.9-B.13) to facilitate comparison and help the committee assess the difficulty of creating and applying them to other parts of the CCSP.

The case studies were created to inform the committee's thinking about metrics. A selection is presented here, in draft form, to show how and why the committee developed general metrics for the CCSP (Box 5.1). No attempt was made to revise the case study metrics after the general metrics were created. The emphasis is on presenting the committee's thought process, not on recommending specific metrics for CCSP program elements.

CASE STUDY THEMES

The committee derived eight key themes from the milestones, products, and payoffs within the CCSP Strategic Plan and developed one or two case studies for each. These themes also conform to the conventional sequence of scientific investigation, starting with the development of new or better observations and ending with improved use of information to advance knowledge or better serve decision making. The themes are:

1. improve data sets in space and time;
2. improve estimates of physical quantities;
3. improve understanding of processes;
4. improve representation of processes;
5. improve assessment of uncertainty, predictability, or predictive capabilities;
6. improve synthesis and assessment to inform;
7. improve the assessment and management of risk; and
8. improve decision support for adaptive management and policy making.

Case study examples of themes 3 and 8 appear in Chapter 5.

Theme 1: Improve Data Sets in Space and Time

Solar Forcing of Climate

Related CCSP Questions, Milestones, and Products. Question 4.1.5: "To what extent are climate changes as observed in instrumental and paleoclimate records related to volcanic and solar variability, and what mechanisms are involved in producing climate responses to these natural forcings?"[1]

Rationale. Understanding how human activities are altering the Earth's climate requires an understanding of the role of natural variability in climate forcing. Therefore, it is essential to know how the Sun's energy output varies and how these variations affect the Earth's climate.

Background. Nine independent satellite measurements of total solar irradiance (TSI) have been made since 1978. These data show that the TSI has changed during recent 11-year solar cycles with 0.1 percent amplitude (Figure B.1). However, the lack of overlapping instruments having in-flight sensitivity tracking precludes detection of any long-term variations of the Sun's TSI on climate time scales, if any are present. The construction of a long-term irradiance composite depends crucially on assumptions made about the degradation of radiometers that lack in-flight tracking capability. Different assumptions yield two different time series. For example, note the two different trends in the energy input at solar minimum in Figure B.1.

---

[1]Climate Change Science Program and Subcommittee on Global Change Research, 2003, *Strategic Plan for the U.S. Climate Change Science Program*, Washington, D.C., p. 43.

TABLE B.1 Example Metrics for Case Study on Solar Forcing of Climate

Type	Example Metrics
Process	• Is there a plan for continuous measurement of other climate variables related to solar irradiance to enable discernment and quantification of the physical, chemical, and biological links between solar irradiance changes and climate? • Is a plan for periodic five-year review of solar measurements available that includes the following: — Are the measurements being made with sufficient precision and accuracy? — Are the measurement plans robust with respect to the requirements for continuity and/or calibration?
Input	• Are the instruments and platforms required for deployment of a TSI measurement system available? • Are the measurements to be made by these instruments relatable to those made using previous technologies? • Yearly reviews of the following: — Sufficient commitment of resources to allow the planned program to be carried out — Sufficient resources being devoted to the development of climate models to utilize the solar measurements properly • Does the best scientific evidence indicate that the resources being devoted to the solar radiation measurements are appropriate, given our need to understand the climate record and predict future climate changes?
Output	• Publication of a peer-reviewed, multiyear record of TSI that is relatable to existing records • Documented, published records of how solar variability has contributed directly and indirectly to past climate change • Quantitative links between measures of solar activity (e.g., sunspot number, solar wind) and solar irradiance at the top of the Earth's atmosphere
Outcome	• Improved ability to forecast non-irradiance-related effects of solar activity • Forecasts of future solar variability and predictions of its climate effect are available for comparison with other climate drivers to determine the nature of climate change • Recognition of direct and indirect mechanisms by which solar variations can influence climate
Impact	• Public understanding of the importance of solar variation in climate change relative to other radiative forcing (e.g., greenhouse gases) is improved

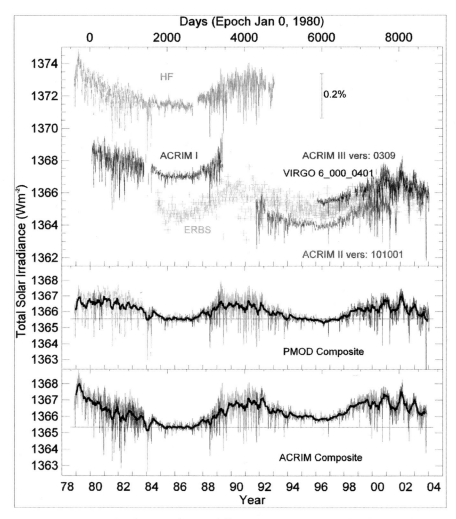

FIGURE B.1 TSI database and two different composite records showing a time series (1978-2004) of measured solar energy input per unit area to the Earth system from various instruments: Active Cavity Radiometer Irradiance Monitor (ACRIM), Variability of Solar Irradiance and Gravity Oscillations (VIRGO), Hickey-Frieden radiometer (HF), Earth Radiation Budget Satellite (ERBS), and Physikalisch-Meteorologisches Observatorium Davos (PMOD). An overlap in the middle of the record between ACRIM I and ACRIM II or use of a technique with absolute calibration would have made it possible to determine whether there is a trend in TSI at solar minimum. SOURCE: Fröhlich, C., and J. Lean, 2004, Solar radiative output and its variations: Evidence and mechanism, *Astronomy and Astrophysics Reviews*, **12**, 273–320. Copyright 2004; used with kind permission of Springer Science and Business Media.

Aerosols and Their Role in Climate Forcing

Related CCSP Questions, Milestones, and Products. Question 3.1: "What are the climate-relevant chemical, microphysical, and optical properties, and spatial and temporal distributions, of human-caused and naturally occurring aerosols?"[2] Milestones, products, and payoffs include (1) improved description of the global distributions of aerosols and their properties; (2) empirically tested evaluation of the capabilities of current models to link emissions to (a) global aerosol distributions and (b) the chemical and radiative properties (and their uncertainties) of aerosols; and (3) better estimates of the radiative forcing of climate change for different aerosol types and the uncertainties associated with those estimates.[3]

Rationale. One of the largest uncertainties in climate research is the specification of aerosol properties and their role in direct climate forcing (Figure B.2). The challenge is to adequately characterize the nature and occurrence of atmospheric aerosols and include their effects in models to reduce uncertainties in climate prediction.

Background. Because aerosols (1) originate from a variety of sources, (2) are distributed across a wide spectrum of particle sizes, and (3) have atmospheric lifetimes that are much shorter than those of most greenhouse gases, their concentrations and composition have great spatial and temporal variability. Satellite-based measurements of aerosols are necessary but not sufficient for acquiring an adequate information base upon which progress in understanding the role of aerosols in climate can be built. In situ measurements and process-level studies are necessary to reduce uncertainties in both direct and indirect forcing. The CCSP strategic plan calls for expanded use of "space-based, airborne, and ground-based instruments and laboratory studies to provide better data for aerosols . . . ," particularly to improve knowledge of spatial distribution and temporal variation of aerosols and precursor gases and of physical, chemical, and optical processes of aerosols; and to distinguish natural from anthropogenic aerosol.[4] These objectives illustrate the need for basic science information to assess the net radiative effect of aerosols.

---

[2]Climate Change Science Program and Subcommittee on Global Change Research, 2003, *Strategic Plan for the U.S. Climate Change Science Program*, Washington, D.C., p. 29.

[3]Climate Change Science Program and Subcommittee on Global Change Research, 2003, *Strategic Plan for the U.S. Climate Change Science Program*, Washington, D.C., p. 33.

[4]Climate Change Science Program and Subcommittee on Global Change Research, 2003, *Strategic Plan for the U.S. Climate Change Science Program*, Washington, D.C., p. 32.

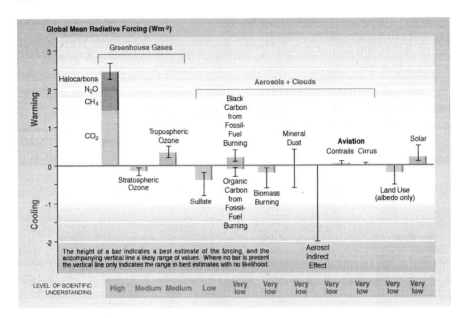

FIGURE B.2 Global annual mean radiative forcing due to a number of agents for the period from preindustrial times (1750) to the present (about 2000). Most of the forcing estimates associated with aerosols have a very low level of scientific understanding, making their estimates highly uncertain. SOURCE: Climate Change Science Program and the Subcommittee on Global Change Research, 2003, *Strategic Plan for the U.S. Climate Change Science Program*, Washington, D.C., p. 16; adapted from Intergovernmental Panel on Climate Change, Working Group I, 2001, *Climate Change 2001: The Scientific Basis*, Cambridge University Press, Cambridge, U.K., p. 392. Used with permission from the Intergovernmental Panel on Climate Change and the Climate Change Science Program.

TABLE B.2 Example Metrics for Case Study on Aerosols and Their Role in Climate Forcing

Type	Example Metrics
Process	• Does a structure exist for the science community to evaluate the adequacy of existing and planned measurement programs concerned with aerosol distribution and radiative properties? • Is there a peer-reviewed five-year plan, updatable every five years, describing where and how measurements will be carried out that link aerosol distribution and chemistry to direct and indirect radiative forcing? • Are the requirements defined for quantifying spatial and temporal variability in planned missions? • Is a mechanism in place to take account of any surprises or new insights in the planning of new measurement campaigns?

TABLE B.2 Continued

Type	Example Metrics
Input	• To what extent do measurements have sufficient accuracy, precision, and completeness to answer the high-priority questions on aerosols and climate? • What resources are being devoted to these measurements? • Are the resources being expended on climate science research being allocated in an optimal manner (i.e., measurements versus models, space measurements versus surface or airborne measurements)? • Does the best scientific evidence indicate that the resources being devoted to solar radiation measurements are appropriate, given our need to understand the climate record and predict future climate changes?
Output	• Well-described and demonstrated relationships between aerosol distribution and radiative forcing • Forecasts of future aerosol distribution and consequences for regional climate based on scenarios of future aerosol emissions
Outcome	• To what extent are the measurements being used to answer the high-priority climate questions that motivated them? • Are the aerosol measurements together with other aerosol research resulting in better understanding of the uncertainties in climate projections due to direct and indirect aerosol processes? • The program leads to regulation of aerosol emissions
Impact	• Regional air quality is improved as a result of aerosol emission regulations

Theme 2: Improve Estimates of Physical Quantities

Sea-Level Rise

Related CCSP Questions, Milestones, and Products. Question 4.2.3: "What are the projected contributions from different components of the climate system to future sea-level changes, what are the uncertainties in the projections, and how can they be reduced?"[5] The associated research activity is "improved representation of processes (e.g., thermal expansion, ice sheets, water storage, coastal subsidence) in climate models that are required for simulating and projecting sea-level changes."[6]

Rationale. Intergovernmental Panel on Climate Change (IPCC) assessments have highlighted considerable uncertainty about the causes of the sea-level rise over the past century. A number of factors can contribute to

---

[5]Climate Change Science Program and Subcommittee on Global Change Research, 2003, *Strategic Plan for the U.S. Climate Change Science Program*, Washington, D.C., pp. 45–46.
[6]Climate Change Science Program and Subcommittee on Global Change Research, 2003, *Strategic Plan for the U.S. Climate Change Science Program*, Washington, D.C., p. 46.

sea-level rise, including ocean thermal expansion, melting of permanent snow cover and mountain glaciers, decreases in groundwater storage, and decreased volume of polar ice sheets. The contributions of ocean thermal expansion are the best constrained, but there is considerable uncertainty about the contributions from groundwater and the Greenland ice sheet. Even the sign of the contribution of the largest freshwater reservoir, the Antarctic ice sheet, is unknown.

Background. The magnitude of sea-level change over the past 100 to 150 years is reasonably well known, owing to a number of observations around the Earth. However, stations give sometimes conflicting measurements, and it is necessary to track changes regionally and over shorter time scales. Integration of measurements with models is essential to estimate the volume of freshwater reservoirs, to determine how this has changed, and to project future changes. Improved estimates of physical quantities are also implicitly required to improve models. Making progress in this research area will require observations (e.g., sea level, geodetic reference frame), estimates of physical quantities (e.g., ice sheet and glacier volume), integration of historical and new information, and improved models to predict sea-level change. The ice sheet volume can be estimated from a number of individual ground (e.g., snow accumulation, ice flow) and satellite-based (e.g., altimetry) measurements, as well as from models relating volume and elevation change.

TABLE B.3 Example Metrics for Case Study on Sea-Level Rise

Type	Example Metrics
Process	• Is there a coordinated, strategic plan that the agencies use to guide research programs, set priorities, and support budget requests? Is the plan responsive to decision support needs?
	• Is there a coordinated, global strategic plan for measurement systems that agencies use to guide new investments, justify ongoing networks, and support budget requests?
	• Do the plan and the program have an appropriate balance of in situ and space-based measurements? Are they well integrated?
Input	• Are there adequate, well-performing data and information systems?
	• Is the research taking advantage of emerging technology and system integration and stimulating the development of new measurement technologies?
Output	• How has the accuracy of measuring sea level and other priority global fluxes and reservoirs of water significantly improved as a result of the deployment of measurement systems for research?
	• Are the measurements of sufficient accuracy to inform assessments and policy?

TABLE B.3 Continued

Type	Example Metrics
	• Have adequate means of assessing measurement accuracy at the scales of interest been developed? • Are research programs producing synthesized results addressing the components of sea-level rise?
Outcome	• Are the research results leading to lower uncertainties in the historical contributions to sea-level rise and thence to better projections of future sea-level rise? • Has significant progress been made on understanding the contributions to sea-level rise as a result of the measurement, process research, and modeling programs? • Do these projections adequately inform assessments and provide a basis for adaptive management and (inter)national policy making on mitigating the potential consequences of sea-level rise (e.g., impacts on coastal communities and ecosystems)?
Impact	• "No-build" zones established between structures (e.g., roads, railways, houses) and the shoreline protect communities from sea-level rise

Theme 4: Improve Representation of Processes

Climate-Vegetation Feedbacks

Related CCSP Questions, Milestones, and Products. Question 8.1: "What are the most important feedbacks between ecological systems and global change (especially climate), and what are their quantitative relationships?"[7] The CCSP product related directly to the improved representation of processes in models is, "Quantification of important feedbacks from ecological systems to climate and atmospheric composition to improve the accuracy of climate projections. This product will be needed to ensure inclusion of appropriate ecological components in future climate models."[8]

Rationale. Our lack of knowledge about the nature of climate-vegetation interactions has hindered our ability to predict climate sensitivity and to understand the response of ecosystems to climate change.

Background. Early studies of tropical deforestation called attention to the importance of vegetation in governing surface energy and moisture

[7]Climate Change Science Program and Subcommittee on Global Change Research, 2003, *Strategic Plan for the U.S. Climate Change Science Program*, Washington, D.C., p. 84.
[8]Climate Change Science Program and Subcommittee on Global Change Research, 2003, *Strategic Plan for the U.S. Climate Change Science Program*, Washington, D.C., p. 86.

fluxes.[9] Climate, in turn, has a significant impact on vegetation. The importance of these effects prompted the development of a wide variety of climate models designed to include atmosphere-vegetation interactions, beginning with Dickinson (1984) and Dorman and Sellers (1989).[10] However, climate-vegetation models are still in their infancy. A better understanding of the controls on vegetation distribution and character, including weather, climate, and the role of human activities (e.g., change in land use and land cover, creation of pollutants), is required to improve predictions of future vegetation distributions. We need a more explicit understanding of the complex interactions between diverse ecosystems and ecosystem components and their chemical as well as physical environment. An improved assessment of the importance of spatial and temporal variability in ecosystem character and an ability to address multiple spatial scales will also be required if we are to quantify changes that will influence moisture and energy budgets. All of these factors require improved field and controlled-environment facilities and long-term observing sites to quantify these interactions. An improved representation of these processes is the key to improved climate-vegetation models. In addition, opportunities to validate the models, perhaps through vegetation records from past climates, will be required if we are to gain confidence in the model predictions.

---

[9]Dickinson, R.E., and A. Henderson-Sellers, 1988, Modelling tropical deforestation: A study of GCM land-surface parameterizations, *Quarterly Journal of the Royal Meteorological Society*, **114**, 439–462; Henderson-Sellers, A., and V. Gornitz, 1984, Possible climatic impacts of land cover transformations, with particular emphasis on tropical deforestation, *Climatic Change*, **6**, 231–257.

[10]Dickinson, R.E., 1984, Modelling evapotranspiration for three-dimensional global climate models, in *Climate Processes and Climate Sensitivity*, J.E. Hansen and T. Takahashi, eds., Geophysical Monograph 29, Maurice Ewing Volume 5, American Geophysical Union, Washington, D.C., pp. 58–72; Dorman, J.L., and P.J. Sellers, 1989, A global climatology of albedo, roughness length and stomatal resistance for atmospheric general circulation models as represented by the simple biosphere model (SiB), *Journal of Applied Meteorology*, **28**, 833–855.

TABLE B.4 Example Metrics for Case Study on Climate-Vegetation Feedbacks

Type	Example Metrics

Process	• Is a functioning peer review process in place involving scientists, managers, and other stakeholders? Are there timetables for periodic peer review of results?
	• Recognized leadership that enables interaction between diverse communities of scientists
	• A five-year plan, revisited every five years, to assess progress and set priorities through peer review for the following:
	  — Implementation of experiments, analysis, and modeling designed to increase understanding of and confidence in the linkages between vegetation and environmental change
	  — Implementation of experiments, analysis, and modeling designed to improve prediction of climate change and variability at a regional level with the resolution and accuracy needed for vegetation studies
	  — Development of field and controlled-environment facilities and long-term ecological observing stations designed to improve understanding and quantification of vegetation-climate interactions
	• An ability to revisit the planning process in response to the development of new experimental methods and new insights from other experiments and fields of study
	• Are systems in place that will promote interaction, partnership, and communication between the ecosystem community and the climate and environmental research community, including scientists, agency managers, policy makers, and the public?

Input	• Sufficient intellectual foundation in multiple disciplines to support the research
	• Available funds for the development and maintenance of a sustainable scientific community and for promoting interaction, partnership, and communication between scientists, agency managers, policy makers, and the public
	• Annual research and development (R&D) expenditures are sufficient to implement and sustain the following:
	  — Principal Investigators (PIs) and/or "centers" projects directed toward achieving the objectives
	  — The investigation of
	    ◦ Competing ideas and interpretations of relationships between climate and vegetation
	    ◦ Innovative and comprehensive approaches for gathering or interpreting and modeling climate-vegetation interactions
	    ◦ The full breadth of relationships between environmental disturbance and ecosystems, including climate, pollutants, and land cover or land use
	    ◦ The resilience of ecosystems to environmental stress
	  — Interpretive activities
	  — Development of environmentally controlled facilities and long-term observing sites
	  — Development of predictive models and synthesis of information
	  — Integration of diverse research communities and existing research enterprises

continued

TABLE B.4 Continued

Type	Example Metrics
Output	• Experimental and observational data of sufficient quantity and quality to support the determination of climate-vegetation relationships • Well-described and demonstrated relationships between environment and vegetation • Climate and climate variability forecasts suitable for determining the future distribution of vegetation, with well-described sources of error and limitations • Vegetation character and distribution projections suitable for determining the impact of vegetation changes on climate • Published reports supporting the analysis of vegetation and climate relationships • Effectively selected, sufficiently accurate, peer-reviewed, published, and broadly accepted data and analysis on vegetation and environment relationships • Adequate community and infrastructure have been developed to support a program of monitoring, surveillance, and modeling of ecosystems • Periodic assessments of the state of the science • Well-described and demonstrated assessment of vegetation-climate interactions
Outcome	• Consistent and reliable projections of vegetation change and climate-vegetation interactions and feedbacks, with well-described sources of error and limitations • Well-described and demonstrated assessment of the resilience of vegetation to a variety of environmental stresses • An improved understanding of the response of ecosystems to environmental stress through an improved capability to assess the role of climate change on a variety of time scales • A peer-reviewed, published, broadly accepted conclusion on the relationships between environment and vegetation • Accelerated incorporation of improved knowledge of climate-vegetation processes and feedbacks into climate models to reduce uncertainty in projections of climate sensitivity and changes in climate and related conditions • Observations, analysis, and models are utilized to improve our understanding of vegetation changes and other ecosystem responses • Expansion of the monitoring, surveillance, and forecast knowledge gained through an examination of vegetation to other areas of ecosystem analysis • Integration of a sustainable community of climate and ecosystem scientists
Impact	• Increased public understanding of the role of climate and other environmental stresses on ecosystems • Evidence of improved ecosystem management as a result of use of improved data and analysis tools and understanding of ecosystem function

Theme 5: Improve Assessment of Uncertainty, Predictability, or Predictive Capabilities

Paleoclimate Time Series as Benchmarks of Climate Variability and Change

Related CCSP Questions, Milestones, and Products. Question 4.1.5: "To what extent are climate changes as observed in instrumental and paleoclimate records related to volcanic and solar variability, and what mechanisms are involved in producing climate responses to these natural forcings?"[11] The related milestone is "targeted paleoclimatic time series as needed, for example, to establish key time series of observations and natural forcing mechanisms as benchmarks of climate variability."[12]

Rationale. Improving our ability to predict climate change, understanding the limits to our ability to make such predictions, and defining our confidence in climate predictions are key elements of the CCSP strategic plan. A critical subset of research directed at these goals includes (1) establishment of time series of prehistorical climate variations, (2) simulation of past climates, the annual mean climate, the seasonal cycle, or seasonal-to-decadal variability (e.g., El Niño-Southern Oscillation), and (3) determining the extent to which physical variables or phenomena can be portrayed realistically in models.[13]

Background. A significant number of studies have used estimates of past radiative forcing (e.g., volcanic eruptions, solar variability) to simulate the climate record of the last several centuries and then compared these model simulations to proxy records of climate change derived from a variety of sources.[14] These studies use different reconstructions of past temperature trends and different model parameters. Nevertheless, model simulations and climate reconstructions, which have independent uncertainties, are generally consistent, suggesting some level of capability in understanding and simulating the climates of the last 1000 years. The success of proxies in defining past climate change depends on the degree to which the

---

[11]Climate Change Science Program and Subcommittee on Global Change Research, 2003, *Strategic Plan for the U.S. Climate Change Science Program*, Washington, D.C., p. 43.

[12]Climate Change Science Program and Subcommittee on Global Change Research, 2003, *Strategic Plan for the U.S. Climate Change Science Program*, Washington, D.C., p. 44.

[13]See associated products identified in Climate Change Science Program and Subcommittee on Global Change Research, 2003, *Strategic Plan for the U.S. Climate Change Science Program*, Washington, D.C., pp. 43–50.

[14]See the review in National Research Council, 2005, *Radiative Forcing of Climate Change: Expanding the Concept and Addressing Uncertainties*, The National Academies Press, Washington, D.C., 222 pp.

proxy accurately reflects temperature, precipitation, atmospheric circulation, and ocean circulation and on the degree to which spatially limited samples reflect hemispheric averages. Spatial patterns of climate change are also of importance, particularly in assessing the capabilities of climate models to simulate the character of past climates beyond hemispheric averages. The near absence of proxy data from large areas of the world presents significant difficulties.[15] There are also significant differences among model simulations of the past, related to differences in model sensitivities or parameterizations and to differences in the specification of past radiative forcing. Resolution of these differences and the development of increasingly credible time series from paleoclimate proxies as benchmarks of climate variability and change are important objectives for the CCSP.

TABLE B.5 Example Metrics for Case Study on Paleoclimate Time Series as Benchmarks of Climate Variability and Change

Type	Example Metrics
Process	• Does a structure exist for scientific community planning and peer review of paleoclimate variability and benchmarking? • Are there processes and timetables for periodic peer review of results generated for each paleoclimate proxy and of synthesis activities that cross or employ multiple proxies and consider different estimates of past radiative forcing? — Does the review enable determination of the comparability and continuity of data generated for different proxies? — Does the review enable testing of model predictions (benchmarks) outside of experimental areas? • A five-year plan for implementation of experiments, analysis, and modeling to obtain an increased understanding of and confidence in the causes of recent and historical climate change, revisited every five years, to assess progress through peer review. It is particularly important that the planning process be revisited periodically in the light of development of new experimental methods and new insights from other experiments and fields of investigation
Input	• Annual R&D expenditures are sufficient to implement and sustain — PIs and/or "centers" projects directed toward achieving the objectives — The investigation of ○ Competing ideas and interpretations of proxy data ○ Innovative approaches for gathering or interpreting paleoclimate records ○ The full breadth of proxy types ○ The application of climate models with estimates of past radiative forcing — Interpretive activities

---

[15]For reference, a comparison of different reconstructions is provided by Jones, P.D., and M.E. Mann, 2004, Climate over past millennia, *Reviews of Geophysics*, **42**, doi: 10.1029/2003RG000143.

TABLE B.5 Continued

Type	Example Metrics
	• Funds available for the development and maintenance of a sustainable paleoclimate scientific community of sufficient depth and diversity
	• Do data of sufficient quantity and quality exist to support the analysis of historical (paleolithic) patterns of climate variability and change?
Output	• Well-described and demonstrated relationships between the observations and model output
	• Description of the potential errors and sources of limitations in the observations, forcing factors, and model capability
	• Improved description of aerosol distribution, solar variability, and land-use or land-cover forcing factors
	• Effectively selected, sufficiently accurate, peer-reviewed, published, and broadly accepted data and analysis on our ability to simulate the climate of the last 1000 years
	• Extension of model-data comparisons for the last 1000 years to the following:
	— Additional variables beyond globally averaged, mean annual surface temperature
	— The spatial and temporal character of climate variability
Outcome	• An improved ability to separate the contributions of natural versus human-induced climate forcing to climate variations and change
	• A peer-reviewed, published, broadly accepted conclusion on our ability to simulate the climate of the last 1000 years, to attribute these variations to specific causes, and to predict future climate
Impact	• Public is better educated on the history of climate change

Theme 6: Improve Synthesis and Assessment to Inform

Human Health and Climate

Related CCSP Questions, Milestones, and Products. Question 9.4: "What are the potential human health effects of global environmental change, and what climate, socioeconomic, and environmental information is needed to assess the cumulative risk to health from these effects?"[16]

Rationale. Synthesis and assessment is a major goal of the CCSP strategic plan. The plan defines assessment as "processes that involve analyzing and evaluating the state of scientific knowledge (and the associated degree of scientific certainty) and, in interaction with users, developing

---

[16]Climate Change Science Program and the Subcommittee on Global Change Research, 2003, *Strategic Plan for the U.S. Climate Change Science Program*, Washington, D.C., p. 98.

information applicable to a particular set of issues or decisions."[17] The impact of climate change on the occurrence of human infectious disease is an example of the importance of synthesis and assessment, given the many possible health effects that can stem from changes in climate and the importance of health to society. Changes in the distribution of disease vectors, for example, may result in exposure of new populations to disease risk and perhaps reexposure of populations to a past disease.

Background. Detecting the effect of climate change on the prevalence and intensity of human infectious disease is difficult because of the uncertainties associated with prediction of long-term changes in climatic conditions, the resultant ecological changes, and their causal connection to specific diseases.[18] Detecting a change in health that can be attributed to climate variability requires defining a baseline distribution for a disease vector prior to climate change. The effect of climate change can be also assessed by changes in the intensity of transmission of diseases, by changes in the transmission season, or by changes in their geographical distribution. The immediate goal is to identify climate processes that appear to affect health directly or to result in environmental changes that facilitate the spread of human diseases. Assessment of the links between climate change and infectious disease will focus on (1) present understanding of the links between weather, climate, and infectious diseases; (2) knowledge of the fundamental relationships between disease and ecological or other environmental conditions (Figure B.3); (3) identification of the factors that result in an improved ability to predict climate conditions (probably at least a season in advance), climate variability, and eventually future climate change; and (4) frameworks for making decisions involving the scale of change, management scope, and stakeholder involvement. Information on these factors is required particularly at the regional level, with the resolution and accuracy appropriate for assessing impacts on ecosystems, assessing risks to human populations, and enabling development of public health strategies.

---

[17]Climate Change Science Program and the Subcommittee on Global Change Research, 2003, *Strategic Plan for the U.S. Climate Change Science Program*, Washington, D.C., p. 194.

[18]Epstein, P.R., 2002, Detecting the infectious disease consequences of climate change and extreme weather events, in *Environmental Change, Climate and Health, Issues and Research Methods*, P. Martens and A.J. McMichael, eds., Cambridge University Press, Cambridge, U.K., pp. 172–196.

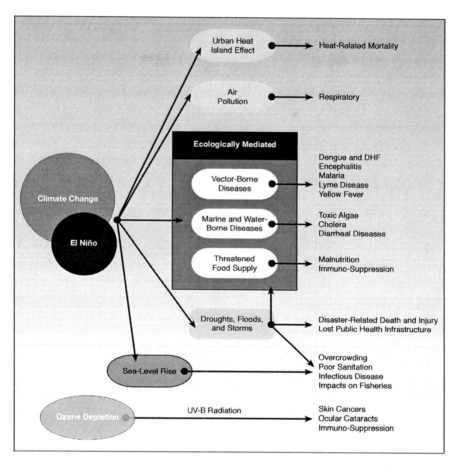

FIGURE B.3 Possible pathways of public health impacts from climate change. SOURCE: Climate Change Science Program and the Subcommittee on Global Change Research, 2003, *Strategic Plan for the U.S. Climate Change Science Program*, Washington, D.C., p. 99; used with permission.

TABLE B.6 Example Metrics for Case Study on Human Health and Climate

Type	Example Metrics
Process	• Is a transparent, inclusive, and peer review process in place for identifying leadership of the assessment activity, structure and timing of the assessment, and selection of participants? • Is there a process for peer review of the assessment and its conclusions, including a process for incorporating reviewer suggestions and comments in the final product? • Is there a process that enables identification of bottlenecks to rapid research progress?
Input	• Does a broad community of professionals and stakeholders required for assessment exist? • Are funds available for the development and maintenance of a sustainable climate and health scientific community of sufficient depth and diversity? • Are funds available for the assessment, including selection of participants, communication among participants and the larger community, preparation, and peer review? • Are funds available for distribution of the assessment and communication of conclusions to a wide audience? • Are historical climate, health, and environmental data available that are of sufficient quantity and quality to support the determination of historical patterns of climate-related health effects?
Output	• Effectively selected, sufficiently accurate, peer-reviewed, published, and broadly accepted data and analysis on health and environment relationships • Climate and climate variability forecasts suitable for assessing health outcomes, with well-described sources of error and limitations • Development of monitoring networks that support forecasting regional-scale climate variability and predicting its impact on human health
Outcome	• Consistent and reliable predictions of climate variables (e.g., sea surface or land temperature distributions) linked to human disease outbreak, with well-described sources of error and limitations • Ability to predict the extent to which a change in climate will significantly affect public health, as measured by an increase in infant mortality rates, declines in human life expectancy, or other factors • Existence of a health care infrastructure with the appropriate expertise to respond to climate predictions
Impact	• Increased public awareness of climate impacts on human health • Predictions of climate change reduce risk of human disease outbreaks

Theme 7: Improve the Assessment and Management of Risk

Assessing, Preventing, and Managing Public Health Threats of Infectious Diseases

Related CCSP Questions, Milestones, and Products. Question 9.3: "How can the methods and capabilities for societal decision-making under conditions of complexity and uncertainty about global environmental variability and change be enhanced?"[19] Question 9.4: "What are the potential human health effects of global environmental change, and what climate, socioeconomic, and environmental information is needed to assess the cumulative risk to health from these effects?"[20] Many milestones and products are related to risk assessment and health. The most direct discussion of risk is found within the human contributions and responses theme: (1) additional tools for preventing and managing the public health threat of infectious diseases; (2) assessments of the potential health effects of combined exposures to climatic and other environmental factors (e.g., air pollution); (3) the next phase of health sector assessments to understand the potential consequences of global change for human health in the United States, especially for at-risk demographic and geographic subpopulations; and (4) improved characterization and understanding of vulnerability and adaptation based on analyses of societal adjustment to climate variability and seasonal-to-interannual forecasts.[21]

Rationale. The last decade has brought much greater emphasis on the potential impacts of climate change on regions and sectors (e.g., forestry, ecosystems, human health, water resources), economic analysis of potential impacts, and making climate information useful to a variety of stakeholders. Consequently, improving our ability to assess and manage risks associated with climate variability and change has become a key goal of the CCSP (Goal 5). The CCSP proposes to develop improved decision support processes and products to improve the use of knowledge on climate change and its potential impacts.

Background. Outbreaks of many infectious diseases are tied to the seasons, suggesting a link to weather and climate. A growing body of

---

[19]Climate Change Science Program and the Subcommittee on Global Change Research, 2003, *Strategic Plan for the U.S. Climate Change Science Program*, Washington, D.C., p. 97.
[20]Climate Change Science Program and the Subcommittee on Global Change Research, 2003, *Strategic Plan for the U.S. Climate Change Science Program*, Washington, D.C., p. 98.
[21]Climate Change Science Program and the Subcommittee on Global Change Research, 2003, *Strategic Plan for the U.S. Climate Change Science Program*, Washington, D.C., pp. 97–100.

research is confirming the connection between weather, climate, and infectious diseases transmitted by mosquitoes, ticks, and rodents (e.g., Lyme disease, malaria, dengue fever, *Hantavirus*).[22] Such linkages hold the promise that disease outbreaks can be anticipated, that the vulnerability of the population can be assessed, and that beneficial adaptation or mitigation strategies can be put into place that minimize or manage risk. These linkages define a foundation for preventing or managing specific threats. The assessment of risk requires a more complete understanding of these linkages, including the following:

1.   An improvement in the ability to assess the vulnerability of regions (e.g., *Hantavirus* in the Southwest) and populations (e.g., the young, the elderly, those with compromised immune systems, the economically disadvantaged). The assessment objective should be to refine the estimate of risk associated with disease outbreaks.

2.   The ability to identify system thresholds or "breakpoints" that may influence the outbreak and spread of disease.

3.   An improved understanding of time lags between climate conditions and the outbreak and spread of diseases.

4.   The enhancement in knowledge of the potential for adaptation and the resilience of communities, institutions, public health care systems, regions, and sectors, including time, cost, efficacy, and feasibility.

TABLE B.7 Example Metrics for Case Study on Assessing, Preventing, and Managing Public Health Threats of Infectious Diseases

Type	Example Metrics
Process	• A five-year plan, revisited every five years, to assess progress and set priorities through peer review, for example: — Implementation of experiments, analysis, and modeling designed to increase understanding of and confidence in the linkages between health and environmental change — Implementation of experiments, analysis, and modeling designed to improve prediction of climate change and variability at a regional level with the resolution and accuracy needed for health studies

[22]National Research Council, 2001, *Grand Challenges in Environmental Sciences*, National Academy Press, Washington, D.C., pp. 36–42; National Research Council, 2001, *Under the Weather: Climate, Ecosystems, and Infectious Disease*, National Academy Press, Washington, D.C., 160 pp.

TABLE B.7 Continued

Type	Example Metrics
	— Development of monitoring and surveillance systems with sufficient data collection frequency, resolution, and accuracy to detect indications of infectious disease outbreaks — Determination of vulnerable regions and populations to define the level of risk associated with disease outbreaks — Determination of the ability of communities, institutions, and public health care systems to respond to disease outbreaks • Are systems are in place that will promote interaction, partnership, and communication between the health community and the climate and environmental research community, including scientists, agency managers, policy makers, and the public? • Are procedures in place for translating model results into knowledge that is understandable and usable by decision makers?
Input	• Are funds available for the development and maintenance of a sustainable scientific community and for promoting interaction, partnership, and communication between scientists, agency managers, policy makers, and the public? • Are annual R&D expenditures sufficient to implement and sustain the following: — PIs and/or "centers" projects directed toward achieving the objectives — The investigation of º Competing ideas and interpretations of relationships between climate and health º Innovative and comprehensive approaches for gathering or interpreting and modeling disease outbreaks º The full breadth of relationships between environmental disturbance and health º Vulnerabilities of human populations º Resilience of communities and institutions — Interpretive activities — Development of a robust disease monitoring and surveillance system — Development of predictive models and synthesis of information • A "climate services" function that enables climate information and predictions to be used by the health community
Output	• Well-described and demonstrated assessment of population vulnerabilities to disease outbreaks • Adequate community and infrastructure have been developed to support a program of monitoring, surveillance, and forecasting of disease risk • Sufficient spatial and temporal coverage of model predictions to provide an adequate disease potential based on monitoring and surveillance, with well-described sources of error and limitations • Effective education mechanisms to promote behavior that will reduce risk
Outcome	• Expansion of the monitoring, surveillance, and forecast knowledge gained through an examination of health to other areas of ecological risk analysis • Consistent and reliable forecasts of disease outbreak potential, with well-described sources of error and limitations • A reliable system for using forecasts to implement adaptation or mitigation strategies that minimize adverse outcomes associated with infectious diseases

continued

TABLE B.7 Continued

Type	Example Metrics
Impact	• Demonstrated cases of successful risk management (e.g., outbreak averted, public warned in time, behavior of individuals changed) • Significantly reduced morbidity and mortality rates as a result of improved management of infectious disease • Improved health infrastructure or heath education programs inform the public of potential risks • Increased public understanding of health risks and requirements to mitigate risk

Theme 8: Improve Decision Support for Adaptive Management and Policy Making

National Policy-Making Case Study: Scenarios of Greenhouse Emissions and Climate Response

Related CCSP Questions, Milestones, and Products. Question 9.2: "What are the current and potential future impacts of global environmental variability and change on human welfare, what factors influence the capacity of human societies to respond to change, and how can resilience be increased and vulnerability decreased?"[23] The relevant payoff emphasizes scenario refinement: "Scenarios will be strengthened by an improved understanding of the interdependence among economic growth; population growth, composition, distribution, and dynamics (including migration); energy consumption in different sectors (e.g., electric power generation, transportation, residential heating and cooling); advancements in technologies; and pollutant emissions."[24]

Rationale. One of the key environmental policy decisions facing the world today is the response to the buildup of greenhouse gas (GHG) emissions. Scenarios—future socioeconomic, technological, or policy developments that may or may not be realized—are a useful way of studying the impacts of GHG emissions on populations and formulating a policy response. Scenario planning is useful when there are considerable uncertainties in some of the future trends and inputs that affect planning.

---

[23]Climate Change Science Program and the Subcommittee on Global Change Research, 2003, *Strategic Plan for the U.S. Climate Change Science Program*, Washington, D.C., p. 95.
[24]Climate Change Science Program and the Subcommittee on Global Change Research, 2003, *Strategic Plan for the U.S. Climate Change Science Program*, Washington, D.C., p. 95.

Background. GHG scenarios include factors such as population growth, economic development, and technological change, and they may also include analysis of the effects of atmospheric GHG concentrations, climate variables, and measures of human and environmental consequences. Such scenarios may or may not assume various levels of emissions control, such as long-term targets for atmospheric stabilization under Article 2 of the Framework Convention on Climate Change.[25] They may take the form of "if-then" questions for choosing a specific policy response, or they may form the basis for policy studies aimed at setting long-term goals or assessing current policy measures, including research and development and other aspects of technology development.

TABLE B.8 Example Metrics for Case Study on Scenarios of Greenhouse Gas Emissions and Climate Response

Type	Example Metrics
Process	• Are effective mechanisms in place for coordination with the Climate Change Technology Program?
	• Does a structure exist for community planning and peer review of scenario development, public policy response, and analysis?
	• Is there a timetable for the periodic review of scenario development activities, including testing of scenarios under different policy approaches?
Input	• Does a program exist that effectively sustains the needed analysis capability?
	• Funds are available for the development and maintenance of a sustainable scientific community capable of analyzing climate change scenarios and policy response
	• Historical climate, health, and environmental data are of sufficient quantity and quality to support the determination of historical patterns of climate-related effects
	• Funds are available to support the technology, monitoring systems, predictive models, and interpretive activities required to develop different climate-related scenarios and to support the assessment of relevant policy responses
Output	• Peer-reviewed results from each region and from cross-region syntheses ensure comparability and continuity of data generated for different regions
	• Have active groups been created that are capable of carrying out the desired policy-related scenario analysis, and is the necessary general analytic capability being sustained to respond when needs arise?
	• Development of scenarios that not only reflect the range of problems produced by climate change, but also—through deliberative processes—are widely acceptable to impacted populations

continued

[25]United Nations, 1992, *United Nations Framework Convention on Climate Change*, New York, 33 pp.

TABLE B.8 Continued

Type	Example Metrics
	• Are the analysis and assessment methods well documented, and is the work published in the peer-reviewed literature?
	• Are the analysis and assessment capabilities adequate to analyze climate scenarios, including the following:
	— An ability to handle multiple gases, multiple sectors, and all regions of the world
	— The capacity to link emissions scenarios to climate outcomes
	— The capability to assess a range of policy proposals and estimate their cost
	— The capability to analyze uncertainty in emissions, climate outcomes, and policy cost
Outcome	• Accepted proposals for domestic emissions control measures
Impact	• Program results are reflected in U.S. government climate policy, international forums (including the IPCC), and/or public discussion of the issue
	• The United States is adequately and appropriately prepared for international climate change negotiations

COMPARISON OF CASE STUDY METRICS

The process, input, output, outcome, and impact metrics developed for the case studies that appear here and in Chapter 5 are grouped in Tables B.9 through B.13 to facilitate comparison and show how the general metrics (Box 6.1) arose.

TABLE B.9 Process Metrics for All Case Studies

Theme	Example Metrics
1	*Solar Forcing*
	• Is there a plan for continuous measurement of other climate variables related to solar irradiance to enable discernment and quantification of the physical, chemical, and biological links between solar irradiance changes and climate?
	• Is a plan for periodic five-year review of solar measurements available that includes the following:
	— Are the measurements being made with sufficient precision and accuracy?
	— Are the measurement plans robust with respect to the requirements for continuity and/or calibration?

TABLE B.9 Continued

Theme	Example Metrics
	Aerosol Forcing • Does a structure exist for the science community to evaluate the adequacy of existing and planned measurement programs concerned with aerosol distribution and radiative properties? • Is there a peer-reviewed five-year plan, updatable every five years, describing where and how measurements will be carried out that link aerosol distribution and chemistry to direct and indirect radiative forcing? • Are the requirements defined for quantifying spatial and temporal variability in planned missions? • Is a mechanism in place to take account of any surprises or new insights in the planning of new measurement campaigns?
2	*Sea-Level Rise* • Is there a coordinated, strategic plan that the agencies use to guide research programs, set priorities, and support budget requests? Is the plan responsive to decision support needs? • Is there a coordinated, global strategic plan for measurement systems that agencies use to guide new investments, justify ongoing networks, and support budget requests? • Do the plan and the program have an appropriate balance of in situ and space-based measurements? Are they well integrated?
3	*Effect of CO_2 on Land Carbon Balance* • Has the leadership of this overall effort, which spans several agencies, been identified? • Does a structure exist that will involve the scientific community in planning of the sites and conditions chosen for manipulation or gradient studies? • Is there a 5-10-year plan for implementation of the manipulation experiments, to be revisited and updated in accord with new discoveries? • Is there a plan to incorporate longer-term aspects of the problem that extend beyond the 5-10-year horizon (i.e., multiple generations of plants exposed to altered atmospheric conditions)? • Do a mechanism and timetable exist for periodic review of experimental implementations, including testing of model predictions outside experimental areas? • Do a mechanism and timetable exist to disseminate results to potential stakeholders (particularly the agricultural community) and involve them in planning discussions?
4	*Climate-Vegetation Feedbacks* • Is a functioning peer review process in place involving scientists, managers, and other stakeholders? Are there timetables for periodic peer review of results? • Recognized leadership that enables interaction between diverse communities of scientists

continued

TABLE B.9 Continued

Theme	Example Metrics

• A five-year plan, revisited every five years, to assess progress and set priorities through peer review for the following:
— Implementation of experiments, analysis, and modeling designed to increase understanding of and confidence in the linkages between vegetation and environmental change
— Implementation of experiments, analysis, and modeling designed to improve prediction of climate change and variability at a regional level with the resolution and accuracy needed for vegetation studies
— Development of field and controlled-environment facilities and long-term ecological observing stations designed to improve understanding and quantification of vegetation-climate interactions
• An ability to revisit the planning process in response to the development of new experimental methods and new insights from other experiments and fields of study
• Are systems in place that will promote interaction, partnership, and communication between the ecosystem community and the climate and environmental research community, including scientists, agency managers, policy makers, and the public?

5   *Paleoclimate Time Series*
• Does a structure exist for scientific community planning and peer review of paleoclimate variability and benchmarking?
• Are there processes and timetables for periodic peer review of results generated for each paleoclimate proxy and of synthesis activities that cross or employ multiple proxies and consider different estimates of past radiative forcing?
— Does the review enable determination of the comparability and continuity of data generated for different proxies?
— Does the review enable testing of model predictions (benchmarks) outside of experimental areas?
• A five-year plan for implementation of experiments, analysis, and modeling to obtain an increased understanding of and confidence in the causes of recent and historical climate change, revisited every five years, to assess progress through peer review. It is particularly important that the planning process be revisited periodically in the light of development of new experimental methods and new insights from other experiments and fields of investigation

6   *Human Health and Climate*
• Is a transparent, inclusive, and peer review process in place for identifying leadership of the assessment activity, structure and timing of the assessment, and selection of participants?
• Is there a process for peer review of the assessment and its conclusions, including a process for incorporating reviewer suggestions and comments in the final product?
• Is there a process that enables identification of bottlenecks to rapid research progress?

TABLE B.9 Continued

Theme	Example Metrics
7	*Assessing, Preventing, and Managing Public Health Threats* • A five-year plan, revisited every five years, to assess progress and set priorities through peer review, for example: — Implementation of experiments, analysis, and modeling designed to increase understanding of and confidence in the linkages between health and environmental change — Implementation of experiments, analysis, and modeling designed to improve prediction of climate change and variability at a regional level with the resolution and accuracy needed for health studies — Development of monitoring and surveillance systems with sufficient data collection frequency, resolution, and accuracy to detect indications of infectious disease outbreaks — Determination of vulnerable regions and populations to define the level of risk associated with disease outbreaks — Determination of the ability of communities, institutions, and public health care systems to respond to disease outbreaks • Are systems are in place that will promote interaction, partnership, and communication between the health community and the climate and environmental research community, including scientists, agency managers, policy makers, and the public? • Are procedures in place for translating model results into knowledge that is understandable and usable by decision makers?
8	*Adaptive Management of Water Resources* • Does the CCSP have an effective planning structure, involving both agency managers and the scientific community, that is used to set priorities and implement water resource programs? • Does an adequate structure exist for peer review of both CCSP water resource programs and the research supported by those programs? • Does the CCSP support programs that effectively sustain research-applications partnerships, carry out a continuing assessment process, and provide test-beds for emerging water resource information and decision support tools? • Is the science in these planned programs responsive to the needs of regional stakeholders? • Does the CCSP water resources plan provide for the measurements, modeling, and decision support needed to link water cycle research and operational needs? *Scenarios of Greenhouse Gas Emissions* • Are effective mechanisms in place for coordination with the Climate Change Technology Program? • Does a structure exist for community planning and peer review of scenario development, public policy response, and analysis? • Is there a timetable for the periodic review of scenario development activities, including testing of scenarios under different policy approaches?

TABLE B.10 Input Metrics for All Case Studies

Theme	Example Metrics
1	*Solar Forcing* • Are the instruments and platforms required for deployment of a TSI measurement system available? • Are the measurements to be made by these instruments relatable to those made using previous technologies? • Yearly reviews of the following: — Sufficient commitment of resources to allow the planned program to be carried out — Sufficient resources being devoted to the development of climate models to utilize the solar measurements properly • Does the best scientific evidence indicate that the resources being devoted to the solar radiation measurements are appropriate, given our need to understand the climate record and predict future climate changes? *Aerosol Forcing* • To what extent do measurements have sufficient accuracy, precision, and completeness to answer the high-priority questions on aerosols and climate? • What resources are being devoted to these measurements? • Are the resources being expended on climate science research being allocated in an optimal manner (i.e., measurements versus models, space measurements versus surface or airborne measurements)? • Does the best scientific evidence indicate that the resources being devoted to solar radiation measurements are appropriate, given our need to understand the climate record and predict future climate changes?
2	*Sea-Level Rise* • Are there adequate, well-performing data and information systems? • Is the research taking advantage of emerging technology and system integration and stimulating the development of new measurement technologies?
3	*Effect of CO_2 on Land Carbon Balance* • Is there sufficient theoretical basis for the design and interpretation of experiments? • Is the technology available to perform experiments assuming multiple, long-term (decadal) manipulations of plots of sufficient size to test hypotheses? • Are sufficient resources (people, dollars) available to implement and support a measurement network, modeling, and interpretive activities for the appropriate period of time (decades)? • Is there an identified stakeholder community to take advantage of scientific advances?
4	*Climate-Vegetation Feedbacks* • Sufficient intellectual foundation in multiple disciplines to support the research

TABLE B.10 Continued

Theme	Example Metrics

- Available funds for the development and maintenance of a sustainable scientific community and for promoting interaction, partnership, and communication between scientists, agency managers, policy makers, and the public
- Annual R&D expenditures are sufficient to implement and sustain the following:
  — PIs and/or "centers" projects directed toward achieving the objectives
  — The investigation of
    ○ Competing ideas and interpretations of relationships between climate and vegetation
    ○ Innovative and comprehensive approaches for gathering or interpreting and modeling climate-vegetation interactions
    ○ The full breadth of relationships between environmental disturbance and ecosystems, including climate, pollutants, and land cover or land use
    ○ The resilience of ecosystems to environmental stress
  — Interpretive activities
  — Development of environmentally controlled facilities and long-term observing sites
  — Development of predictive models and synthesis of information
  — Integration of diverse research communities and existing research enterprises

5   *Paleoclimate Time Series*
- Annual R&D expenditures are sufficient to implement and sustain:
  — PIs and/or "centers" projects directed toward achieving the objectives
  — The investigation of
    ○ Competing ideas and interpretations of proxy data
    ○ Innovative approaches for gathering or interpreting paleoclimate records
    ○ The full breadth of proxy types
    ○ The application of climate models with estimates of past radiative forcing
  — Interpretive activities
- Funds available for the development and maintenance of a sustainable paleoclimate scientific community of sufficient depth and diversity
- Do data of sufficient quantity and quality exist to support the analysis of historical (paleolithic) patterns of climate variability and change?

6   *Human Health and Climate*
- Does a broad community of professionals and stakeholders required for assessment exist?
- Are funds available for the development and maintenance of a sustainable climate and health scientific community of sufficient depth and diversity?
- Are funds available for the assessment, including selection of participants, communication among participants and the larger community, preparation, and peer review?
- Are funds available for distribution of the assessment and communication of conclusions to a wide audience?
- Are historical climate, health, and environmental data available that are of sufficient quantity and quality to support the determination of historical patterns of climate-related health effects?

continued

TABLE B.10 Continued

Theme	Example Metrics
7	*Assessing, Preventing, and Managing Public Health Threats* • Are funds available for the development and maintenance of a sustainable scientific community and for promoting interaction, partnership, and communication between scientists, agency managers, policy makers, and the public? • Are annual R&D expenditures sufficient to implement and sustain the following: — PIs and/or "centers" projects directed toward achieving the objectives — The investigation of ○ Competing ideas and interpretations of relationships between climate and health ○ Innovative and comprehensive approaches for gathering or interpreting and modeling disease outbreaks ○ The full breadth of relationships between environmental disturbance and health ○ Vulnerabilities of human populations ○ Resilience of communities and institutions — Interpretive activities — Development of a robust disease monitoring and surveillance system — Development of predictive models and synthesis of information • A "climate services" function that enables climate information and predictions to be used by the health community
8	*Adaptive Management of Water Resources* • Annual R&D expenditures are sufficient to implement and sustain the following: — PIs and/or "centers" projects directed toward achieving the objectives — The investigation of ○ Competing ideas and interpretations of causes ○ Competing interpretations of data ○ Innovative approaches for gathering or interpreting water resources data • Funds are available for the development and maintenance of a sustainable water resources scientific community of sufficient depth and diversity *Scenarios of Greenhouse Gas Emissions* • Does a program exist that effectively sustains the needed analysis capability? • Funds are available for the development and maintenance of a sustainable scientific community capable of analyzing climate change scenarios and policy response • Historical climate, health, and environmental data are of sufficient quantity and quality to support the determination of historical patterns of climate-related effects • Funds are available to support the technology, monitoring systems, predictive models, and interpretive activities required to develop different climate-related scenarios and to support the assessment of relevant policy responses

TABLE B.11 Output Metrics for All Case Studies

Theme	Example Metrics
1	*Solar Forcing* • Publication of a peer-reviewed, multiyear record of TSI that is relatable to existing records • Documented, published records of how solar variability has contributed directly and indirectly to past climate change • Quantitative links between measures of solar activity (e.g., sunspot number, solar wind) and solar irradiance at the top of the Earth's atmosphere *Aerosol Forcing* • Well-described and demonstrated relationships between aerosol distribution and radiative forcing • Forecasts of future aerosol distribution and consequences for regional climate based on scenarios of future aerosol emissions
2	*Sea-Level Rise* • How has the accuracy of measuring sea level and other priority global fluxes and reservoirs of water significantly improved as a result of the deployment of measurement systems for research? • Are the measurements of sufficient accuracy to inform assessments and policy? • Have adequate means of assessing measurement accuracy at the scales of interest been developed? • Are research programs producing synthesized results addressing the components of sea-level rise?
3	*Effect of CO_2 on Land Carbon Balance* • Peer-reviewed, published results generated for each site and synthesis activities across sites that identify the most important mechanisms at work • Production of a facility that (1) can be put into the field for years at a time and (2) can maintain atmospheric CO_2 levels at a specific set point (e.g., 50 ppm [parts per million] above ambient levels), with a precision (averaged over 1 hour) of 5 ppm. For a subset of these systems, additional control over either atmospheric ozone levels, temperature (i.e., increase by $5°C$ compared to the control plot), soil moisture, or species diversity is required • Development of a suite of new measurement techniques that can detect carbon allocation patterns on time scales of (1) hours, (2) days to weeks, and (3) a growing season in response to external variables and photosynthetic rates of plants in control versus experimentally manipulated systems • Incorporation of relationships between photosynthetic rates, carbon allocation, and external and internal variables into process-based models that simulate patterns of photosynthetic response and allocation (on appropriate time scales for each process) and that can be tested against other observations as well as in other kinds of manipulated systems • Technology developed for rapid control of trace gas concentrations at high precision

continued

TABLE B.11 Continued

Theme	Example Metrics

4   *Climate-Vegetation Feedbacks*
- Experimental and observational data of sufficient quantity and quality to support the determination of climate-vegetation relationships
- Well-described and demonstrated relationships between environment and vegetation
- Climate and climate variability forecasts suitable for determining the future distribution of vegetation, with well-described sources of error and limitations
- Vegetation character and distribution projections suitable for determining the impact of vegetation changes on climate
- Published reports supporting the analysis of vegetation and climate relationships
- Effectively selected, sufficiently accurate, peer-reviewed, published, and broadly accepted data and analysis on vegetation and environment relationships
- Adequate community and infrastructure have been developed to support a program of monitoring, surveillance, and modeling of ecosystems
- Periodic assessments of the state of the science
- Well-described and demonstrated assessment of vegetation-climate interactions

5   *Paleoclimate Time Series*
- Well-described and demonstrated relationships between the observations and model output
- Description of the potential errors and sources of limitations in the observations, forcing factors, and model capability
- Improved description of aerosol distribution, solar variability, and land-use or land-cover forcing factors
- Effectively selected, sufficiently accurate, peer-reviewed, published, and broadly accepted data and analysis on our ability to simulate the climate of the last 1000 years
- Extension of model-data comparisons for the last 1000 years to the following:
  — Additional variables beyond globally averaged, mean annual surface temperature
  — The spatial and temporal character of climate variability

6   *Human Health and Climate*
- Effectively selected, sufficiently accurate, peer-reviewed, published, and broadly accepted data and analysis on health and environment relationships
- Climate and climate variability forecasts suitable for assessing health outcomes, with well-described sources of error and limitations
- Development of monitoring networks that support forecasting regional-scale climate variability and predicting its impact on human health

7   *Assessing, Preventing, and Managing Public Health Threats*
- Well-described and demonstrated assessment of population vulnerabilities to disease outbreaks
- Adequate community and infrastructure have been developed to support a program of monitoring, surveillance, and forecasting of disease risk

TABLE B.11 Continued

Theme	Example Metrics

• Sufficient spatial and temporal coverage of model predictions to provide an adequate disease potential based on monitoring and surveillance, with well-described sources of error and limitations
• Effective education mechanisms to promote behavior that will reduce risk

8   *Adaptive Management of Water Resources*
• Established (accepted, peer-reviewed, published) baselines for hydrologic forecasting improved as a result of CCSP-supported research
• Consistent and reliable estimates and forecasts of water resources quantities (e.g., volume of natural water reservoirs, fluxes) to support adaptive management
• Water resource planning scenarios that take into account contingencies such as substantial decreases in mountain snowpack expected as a result of further climate warming or multiyear droughts that stress water resources systems well beyond their design capacity
• Accurate regional and national measures of the hydrologic effects likely associated with climate change
• Quantitative information on components of the regional, national, and global water cycle that are important for water resources management, such as precipitation patterns and trends, streamflow trends, snowpack, and groundwater changes
• Establishment of the degree to which these components are changing because of factors other than natural variability, such as moisture fluxes and precipitation
• Sustainable information systems that make water resource data and information readily available to research and applications users

Scenarios of Greenhouse Gas Emissions
• Peer-reviewed results from each region and from cross-region syntheses ensure comparability and continuity of data generated for different regions
• Have active groups been created that are capable of carrying out the desired policy-related scenario analysis, and is the necessary general analytic capability being sustained to respond when needs arise?
• Development of scenarios that not only reflect the range of problems produced by climate change, but also—through deliberative processes—are widely acceptable to impacted populations
• Are the analysis and assessment methods well documented, and is the work published in the peer-reviewed literature?
• Are the analysis and assessment capabilities adequate to analyze climate scenarios, including the following:
    — An ability to handle multiple gases, multiple sectors, and all regions of the world
    — The capacity to link emissions scenarios to climate outcomes
    — The capability to assess a range of policy proposals and estimate their cost
    — The capability to analyze uncertainty in emissions, climate outcomes, and policy cost

TABLE B.12 Outcome Metrics for All Case Studies

Theme	Example Metrics
1	*Solar Forcing* • Improved ability to forecast non-irradiance-related effects of solar activity • Forecasts of future solar variability and predictions of its climate effect are available for comparison with other climate drivers to determine the nature of climate change • Recognition of direct and indirect mechanisms by which solar variations can influence climate *Aerosol Forcing* • To what extent are the measurements being used to answer the high-priority climate questions that motivated them? • Are the aerosol measurements together with other aerosol research resulting in better understanding of the uncertainties in climate projections due to direct and indirect aerosol processes? • The program leads to regulation of aerosol emissions
2	*Sea-Level Rise* • Are the research results leading to lower uncertainties in the historical contributions to sea-level rise and thence to better projections of future sea-level rise? • Has significant progress been made on understanding the contributions to sea-level rise as a result of the measurement, process research, and modeling programs? • Do these projections adequately inform assessments and provide a basis for adaptive management and (inter)national policy making on mitigating the potential consequences of sea-level rise (e.g., impacts on coastal communities and ecosystems)?

3	*Effect of CO_2 on Land Carbon Balance*
	• Peer-reviewed and published knowledge of the processes by which increasing atmospheric CO_2 can influence the carbon balance at (1) the whole plant level and (2) the ecosystem level. Determination of the sign and magnitude (to 30%) of the feedback between CO_2 levels and the amount of carbon stored over the first year of the manipulation (and subsequent years as they become available)
	• Models of suitable spatial scale that incorporate process-level understanding are used to predict the response of ecosystems to multiple stressors, such as increased CO_2 and temperature or CO_2 and ozone
	• Policy makers are informed about
	   — The potential for different kinds of ecosystems to store or release carbon under conditions of a 50 ppm increase in atmospheric CO_2
	   — The magnitude of release or uptake of CO_2 and how this understanding will be modified by the presence of more investigators in the field
	• Peer-reviewed assessments that quantify the potential effects of changing atmospheric composition on the yield of different crops
	• Improved prediction of future trends in atmospheric CO_2 levels, given a scenario of fossil fuel emissions and deforestation

TABLE B.12 Continued

Theme	Example Metrics
4	*Climate-Vegetation Feedbacks* • Consistent and reliable projections of vegetation change and climate-vegetation interactions and feedbacks, with well-described sources of error and limitations • Well-described and demonstrated assessment of the resilience of vegetation to a variety of environmental stresses • An improved understanding of the response of ecosystems to environmental stress through an improved capability to assess the role of climate change on a variety of time scales • A peer-reviewed, published, broadly accepted conclusion on the relationships between environment and vegetation • Accelerated incorporation of improved knowledge of climate-vegetation processes and feedbacks into climate models to reduce uncertainty in projections of climate sensitivity and changes in climate and related conditions • Observations, analysis, and models are utilized to improve our understanding of vegetation changes and other ecosystem responses • Expansion of the monitoring, surveillance, and forecast knowledge gained through an examination of vegetation to other areas of ecosystem analysis • Integration of a sustainable community of climate and ecosystem scientists
5	*Paleoclimate Time Series* • An improved ability to separate the contributions of natural versus human-induced climate forcing to climate variations and change • A peer-reviewed, published, broadly accepted conclusion on our ability to simulate the climate of the last 1000 years, to attribute these variations to specific causes, and to predict future climate
6	*Human Health and Climate* • Consistent and reliable predictions of climate variables (e.g., sea surface or land temperature distributions) linked to human disease outbreak, with well-described sources of error and limitations • Ability to predict the extent to which a change in climate will significantly affect public health, as measured by an increase in infant mortality rates, declines in human life expectancy, or other factors • Existence of a health care infrastructure with the appropriate expertise to respond to climate predictions
7	*Assessing, Preventing, and Managing Public Health Threats* • Expansion of the monitoring, surveillance, and forecast knowledge gained through an examination of health to other areas of ecological risk analysis • Consistent and reliable forecasts of disease outbreak potential, with well-described sources of error and limitations • A reliable system for using forecasts to implement adaptation or mitigation strategies that minimize adverse outcomes associated with infectious diseases

continued

TABLE B.12 Continued

Theme	Example Metrics
8	*Adaptive Management of Water Resources* • Effective pilot research-applications partnerships result in experimental use of more accurate hydrologic forecasting tools and improved decision making • A regional demand exists among stakeholders for emerging CCSP data and information to support decision making • Decision support systems have been adapted to use emerging CCSP data and information • Improved information and technology have resulted in improved operational management of water resources, such as water allocations and reservoir operations • New infrastructure (e.g., groundwater backup systems for surface reservoirs) provides a more stable supply of water • More effective water resources planning structures, such as state drought task forces and agency capital investment plans, have been initiated that explicitly consider climate change *Scenarios of Greenhouse Gas Emissions* • Accepted proposals for domestic emissions control measures

TABLE B.13 Impact Metrics for All Case Studies

Theme	Example Metrics
1	*Solar Forcing* • Public understanding of the importance of solar variation in climate change relative to other radiative forcing (e.g., greenhouse gases) is improved *Aerosol Forcing* • Regional air quality is improved as a result of aerosol emission regulations
2	*Sea-Level Rise* • "No-build" zones established between structures (e.g., roads, railways, houses) and the shoreline protect communities from sea-level rise
3	*Effect of CO_2 on Land Carbon Balance* • Crop productivity is improved because of use of forecasts that take into account changes in CO_2, ozone, and climate • Conservation reserves are more resilient because of use of knowledge of how changes in CO_2 affect plant competition and ecosystem structure
4	*Climate-Vegetation Feedbacks* • Increased public understanding of the role of climate and other environmental stresses on ecosystems • Evidence of improved ecosystem management as a result of use of improved data and analysis tools and understanding of ecosystem function

TABLE B.13 Continued

Theme	Example Metrics
5	*Paleoclimate Time Series* • Public is better educated on the history of climate change
6	*Human Health and Climate* • Increased public awareness of climate impacts on human health • Predictions of climate change reduce risk of human disease outbreaks
7	*Assessing, Preventing, and Managing Public Health Threats* • Demonstrated cases of successful risk management (e.g., outbreak averted, public warned in time, behavior of individuals changed) • Significantly reduced morbidity and mortality rates as a result of improved management of infectious disease • Improved health infrastructure or heath education programs inform the public of potential risks • Increased public understanding of health risks and requirements to mitigate risk
8	*Adaptive Management of Water Resources* • Increased resilience of the water supply has decreased the vulnerability of populations to hydrologic aspects of climate variability and change *Scenarios of Greenhouse Gas Emissions* • Program results are reflected in U.S. government climate policy, international forums (including the IPCC), and/or public discussion of the issue • The United States is adequately and appropriately prepared for international climate change negotiations

Appendix C

Pool of Generic Metrics for Science and Technology

A generic set of metrics for evaluating science and technology programs has been developed by E. Geisler, based on a review of the literature.[1] The following metrics are the subset that is most relevant to an agency research and development (R&D) program, and they are categorized according to the committee's definitions of process, input, and output metrics. Geisler did not identify generic outcome or impact metrics.

PROCESS METRICS

Organizational, Strategic, and Managerial Metrics

1. Internal or cycle time: period from the start of a project to transferring an outcome to a downstream unit within the organization.

2. External or commercial cycle time: period from the start of the project to the ultimate sale of a product or service to an external customer.

3. Existence of project champion: number or portion of current projects that have an identifiable champion in the form of a manager from outside the R&D unit.

[1]Geisler, E., 1999, The metrics of technology evolution: Where we stand and where we should go from here, Annual Technology Transfer Society Meeting, July 15-17, 1999, <http://www.stuart.iit.edu/faculty/workingpapers/ technology/>.

4.   Projects with interfunctional teams: number of projects that employ teams composed of people from units across the organization and outside the R&D unit.

5.   Evaluation of the scientific and technical capabilities of the R&D unit and, by extension, of the total organization: external evaluation primarily by various customers, of the capability of the firm and its R&D unit in meeting the scientific and technological challenges of changing markets.

6.   Project progress and success: progress in meeting established objectives and milestones over a given period of time; number or percentage of projects that exhibited technical success on time and on budget.

7.   Evaluation of projects and programs: averages of cost per project, by type of project.

8.   Ownership, support, and funding of projects and programs: percentage of projects supported and funded by other units in the organization that are directly related to a product line or similar commercial entity in the organization; distribution of projects and programs by source of organization.

9.   Human relations measures of R&D personnel: morale of personnel; satisfaction with their work.

10. Relation of R&D to strategic objectives: degree to which R&D objectives are related to the strategic objectives of the organization and are current with any changes in the organization's strategy.

11. Benchmarking project and program performance: relation of project management metrics to benchmarks that are standards, averages, or best practices in the industry or sector; extent to which these benchmarks influence the strategic direction of both R&D and the total organization.

Peer Review Metrics

1.   Internal evaluation: subjective rating by other people in the organization ranked on a scale that measures judgment of respondents.

2.   External evaluation: subjective evaluation by a panel of experts.

3.   Targeted reviews: panel evaluations of any R&D outcome. This may be considered a measure of quality, as viewed by expert reviewers.

INPUT METRICS

Investments in R&D

1.   Expenditures for each stage of research and development.

2.   Expenditures per time frame, for one time period, or over several time periods.

3.   Distribution by categories of expenditures for personnel, equipment, et cetera.

4. Source of funding.

5. Comparison of expenditures, per item category, by competitors, industry averages, and sector averages.

6. Expenditures by discipline, technology, and scientist and engineer.

7. Expenditures related to a product line or other commercial unit of reference, such as customer or market.

OUTPUT METRICS

Bibliometric Measures

1. Publications.

2. Citation analysis.

3. Co-word analysis and database tomography: analyses performed on large databases of R&D bibliographical outcomes, in a form of data mining.

4. Special presentations and honors.

Stages of Outcomes

1. Immediate outputs: proximal or direct outputs from the R&D activity, such as bibliometric measures.

2. Intermediate outputs: outputs of the organizations and entities that have received the immediate outputs, transformed them, and are providing the transformed outputs to other entities in society and the economy.

3. Pre-ultimate outputs: products and services that are generated by those social and economic entities that have received and transformed the intermediate outputs.

4. Ultimate outputs: things of value to the economy and society that were impacted by the pre-ultimate outputs.

5. Index of leading indicators: weighted measures of core and organization-specific measures intended to provide a quantitative appraisal of the value of R&D at each stage of the innovation process.

6. Value indices for leading indicators: value of each index at each stage of the innovation continuum. Value indices are computed by subtracting the value of each leading index from the index that succeeded it. Net value is computed by comparison with costs of R&D and transformation at each stage.

7. Portion of R&D at each stage: the role that R&D has in each of the stages for each of the recipient or transforming organizations. These measures offer a look at the size and value of the R&D contribution for each output, as well as in toto.

Appendix D

Biographical Sketches of
Committee Members

Eric J. Barron (*chair*) is dean of the College of Earth and Mineral Sciences and a distinguished professor of geosciences at the Pennsylvania State University. He led Penn State's Earth System Sciences Center for 15 years and has chaired many committees related to global change, including the National Research Council's (NRC's) Board on Atmospheric Sciences and Climate (BASC), its Climate Research Committee, and the National Aeronautics and Space Administration's (NASA's) Earth Observing System Science Executive Committee. Dr. Barron's research interests are in climate modeling, hydrology, and Earth system history. He is a fellow of the American Geophysical Union and the American Meteorological Society.

Roger C. Bales is a professor of hydrology and water resources in the School of Engineering at the University of California, Merced. His research interests focus on snow hydrology, hydrogeochemistry, water resources, and climate impacts. He was principal investigator of the National Oceanic and Atmospheric Administration's (NOAA's) Climate Assessment for the Southwest Project, which examined the impacts of climate variability and longer-term climate change on human and natural systems in the Southwest. Dr. Bales is a former member of the NRC Committee on Geophysical and Environmental Data and Committee on Hydrologic Studies. He is a fellow of the American Association for the Advancement of Science, the American Geophysical Union, and the American Meteorological Society.

John B. Carberry is director of environmental technology at the DuPont

Company. While his early career focused on developing chemical processes or new products, he is currently analyzing environmental issues of interest to his company to help set policy or develop business programs. In that capacity, he has formulated performance metrics for industrial ecology and presented them in a wide range of venues. He has also participated in a number of global change-related activities, including the mid-Atlantic assessment of the environment. Mr. Carberry has served on a number of committees dealing with performance metrics, most notably the NRC Committee on Industrial Environmental Performance Metrics and the American Institute for Chemical Engineering's Sustainability Metrics Working Group.

David J.C. Constable is director of sustainable development, environment, health and safety product stewardship, corporate environment, health, and safety at GlaxoSmithKline. In addition to his other duties, he is responsible for developing the company's sustainability metrics. He has brought this expertise to the American Institute of Chemical Engineers, where he participated in or led a number of working groups developing sustainability metrics for industrial issues, such as energy, water usage, and pollutants. Dr. Constable also has experience working with government agencies and academia, mostly to advance state-of-the-art environmental technologies.

Paul V. Desanker is an associate professor of geography at the Pennsylvania State University. His research focuses on forest landscape management, the effects of land-use change on ecosystem processes, and the assessment of impacts and of adaptation to climate change. Much of his work concentrates on Africa, and he has served on numerous committees related to climate change on that continent. He is also a member of the United Nation's Framework Convention on Climate Change Least Developed Countries Expert Group and of the Intergovernmental Panel on Climate Change's (IPCC's) Task Group on Climate Impacts Assessments.

Marvin A. Geller is a professor of atmospheric sciences at the State University of New York at Stony Brook. His research deals with atmospheric dynamics, middle and upper atmosphere, climate variability, and aeronomy. Dr. Geller has served on many national and international advisory committees on atmospheric science, the upper atmosphere, and near-space environment and is currently president of the Scientific Committee on Solar-Terrestrial Physics. He is a fellow of the American Meteorological Society and the American Geophysical Union (AGU), and a past president of AGU's Atmospheric Sciences Section.

Eileen E. Hofmann is a professor in the Department of Oceanography at Old Dominion University. Her research focuses on analysis and modeling of biological and physical interactions in marine ecosystems. Dr. Hofmann has served on many ocean-related committees, including the NRC's Ocean Studies Board, and has served as an officer for the Ocean Sciences Section of the American Geophysical Union. She currently chairs the International Global Ocean Ecosystem Dynamics Southern Ocean Planning Group and is member of the Joint Global Ocean Flux Study Synthesis and Modeling Project.

Henry D. Jacoby is a professor of management and co-director of the Joint Program on the Science and Policy of Global Change at the Massachusetts Institute of Technology (MIT). He was formerly director of MIT's Center for Energy and Environmental Policy Research. Dr. Jacoby has made contributions to the study of policy and management in the areas of energy, natural resources, and environment. He has also served on a number of committees related to these topics, including the NRC Climate Impact Committee and the Office of Technology Assessment Committee on Alternative Energy R&D Strategies.

Joyce E. Penner is a professor in the Department of Atmospheric, Oceanic, and Space Sciences and director of the Laboratory for Atmospheric Science and Environmental Research at the University of Michigan. Her research interests focus on cloud and aerosol interactions, interactions of atmospheric chemistry with climate, and model interpretation. Dr. Penner has chaired or been a member of numerous advisory committees related to atmospheric chemistry and global change. Examples include the ad hoc Steering Committee to develop a National Aerosol Climate Interactions Program Plan and the NRC Panel on Aerosol Forcing and Climate Change. She is a fellow of the American Geophysical Union.

Eugene A. Rosa is a professor of sociology and the Edward R. Meyer Distinguished Professor of Natural Resource and Environmental Policy in the Thomas S. Foley Institute for Public Policy and Public Service at Washington State University. His current research focuses on two complementary topics: technological risk and global environmental change. Research activities associated with the latter include specifying the anthropogenic causes of carbon dioxide loads, historical relationships between greenhouse gases and societal well-being, the history of social thought on climate, and theories of environmental impact. Dr. Rosa is a member of the NRC Board on Radioactive Waste Management and the Committee to Review the U.S. Climate Chance Science Program.

Susan E. Trumbore is a professor of Earth system science and director of the Center for Global Environmental Change Research and Institute for Geophysics and Planetary Physics at the University of California, Irvine.

Her research interests are in biogeochemistry and its application to ecology, soil biochemistry, and terrestrial carbon cycling. Dr. Trumbore was an author of the IPCC's report on land use, land-use change, and forestry. She is a fellow of the American Geophysical Union, a former president of its biogeochemistry section, and has served on AGU Committees on Global Environmental Change and Paleoceanography and Paleoclimatology.

Karl K. Turekian is Sterling Professor of Geology and Geophysics and director (until January 1, 2004) of the Institute for Biospheric Studies at Yale University. He is also director of the Center for the Study of Global Change of that institute. His research focuses on applications of isotope geochemistry to marine, atmosphere, terrestrial, and hydrologic environments. He also has a long-standing interest in climate change and has been a member of many NRC committees concerned with that topic. Recent examples include the Committee on Global Change Research, the Ocean Studies Board, and the Water Science and Technology Board. Dr. Turekian is a member of the National Academy of Sciences.

Carl Wunsch is Cecil and Ida Green Professor of Physical Oceanography at the Massachusetts Institute of Technology. His research focuses on ocean observing technologies, and the general circulation of the ocean and its implications for climate change. Dr. Wunsch has chaired a number of ocean science advisory groups, such as the NRC Ocean Studies Board and the International Steering Group for the World Ocean Circulation Experiment. He is a member of the National Academy of Sciences, a foreign member of the Royal Society, a recipient of the American Geophysical Union's Macelwane Award and Maurice Ewing Medal, and the American Meteorological Society's Henry Stommel Medal.

NRC Staff

Anne M. Linn, senior program officer, has been with the NRC Board on Earth Sciences and Resources since 1993. In addition to staffing a wide variety of studies on geophysics, Earth observing systems, and data policy, she directs the U.S. World Data Center Coordination Office. She is also the secretary of the International Council for Science (ICSU) Panel on World Data Centers. Prior to joining the staff of the National Academies, Dr. Linn was a visiting scientist at the Carnegie Institution of Washington and a postdoctoral geochemist at the University of California, Berkeley. She holds a Ph.D. in geology from the University of California, Los Angeles, and an M.S. and B.S. in geology from Texas A&M University.

Appendix E

Abbreviations and Acronyms

ACRIM	Active Cavity Radiometer Irradiance Monitor
AIChE	American Institute of Chemical Engineers
CCSP	Climate Change Science Program
CFC	chlorofluorocarbon
DOE	Department of Energy
ENSO	El Niño-Southern Oscillation
EPA	Environmental Protection Agency
ERBS	Earth Radiation Budget Satellite
FY	fiscal year
GHG	greenhouse gas
GPRA	Government Performance and Results Act
HF	Hickey-Frieden radiometer
IPCC	Intergovernmental Panel on Climate Change
NASA	National Aeronautics and Space Administration
NOAA	National Oceanic and Atmospheric Administration
NRC	National Research Council

NSF   National Science Foundation

OMB   Office of Management and Budget

PART   Program Assessment Rating Tool
PI   principal investigator
PMOD   Physikalisch-Meteorologisches Observatorium Davos

R&D   research and development
RISA   Regional Integrated Science and Assessments

TAO   Tropical Atmosphere Ocean
TOGA   Tropical Ocean Global Atmosphere
TSI   total solar irradiance

USGCRP   U.S. Global Change Research Program
USGS   U.S. Geological Survey
UV   ultraviolet

VIRGO   Variability of Solar Irradiance and Gravity Oscillations

WREN   Washington Research Evaluation Network